ASIA–PACIFIC CLIMATE REPORT 2024

CATALYZING FINANCE AND POLICY SOLUTIONS

AN DEVELOPMENT BANK

 Creative Commons Attribution 3.0 IGO license (CC BY 3.0 IGO)

 Printed on recycled paper

CONTENTS

Tables, Figures, and Boxes   v

Foreword   viii

Acknowledgments   ix

Definitions   xi

Abbreviations   xii

Highlights   xiv

Executive Summary   xvii

1   Addressing the Climate Crisis in Asia and the Pacific   1

1.1   The Need for Accelerated Climate Action   2
1.2   Opportunities Arising from Low-Carbon Transition   3
1.3   Seizing the Moment for Climate Action   5
1.4   Challenges for Developing Asia   5
1.5   The Need for Adaptation   10
1.6   Scope for Higher Ambitions and More Concrete Plans   11
1.7   Supporting a Just, Inclusive, and Comprehensive Transition   17
References   19

2   Impacts and Costs of Climate Inaction for Asia and the Pacific   21

2.1   A Climate in Crisis   22
2.2   Climate Futures for Asia and the Pacific   24
2.3   Climate Change Impacts on Vulnerable Sectors and Populations   33
2.4   Economic Impacts of Climate Change   42
Annex: Details of Modeled Losses of Gross Domestic Product Under a High End Emissions Scenario   49
References   51

3   Accelerating Climate Change Adaptation in Asia and the Pacific   55

3.1   Framing Adaptation Responses to Climate Change   56
3.2   Adaptation Needs and Opportunities   60
3.3   Adaptation Policy Responses to Date   75
3.4   Policy Recommendations to Enhance Adaptation Outcomes   83
References   85

4 Scaling Up Private Climate Capital for Mitigation and Adaptation   91

4.1 Shortfalls in Climate Finance for Asia and the Pacific   92
4.2 The Landscape of Private Climate Capital in Asia and the Pacific   92
4.3 Drivers of Private Climate Investment   95
4.4 Addressing Impediments to Mobilizing Private Climate Capital   109
4.5 Global Initiatives and the Role of Multilateral Organizations   123
References   124

5 Carbon Pricing as a Key Policy for Climate Change Mitigation   131

5.1 Mounting Evidence in Support of Carbon Pricing   132
5.2 Fossil Fuel Subsidies as Negative Carbon Pricing   133
5.3 Carbon Taxation as a Core Carbon Pricing Policy   139
5.4 Build and Connect Emissions Trading Systems   143
5.5 The Role of International Carbon Markets   152
5.6 The Effects of Carbon Pricing on Developing Economies   156
References   162

TABLES, FIGURES, AND BOXES

Tables

1.1 Beliefs, Concerns, and Support for Policies on Climate Change 3
1.2 Select Key Milestones to 2070 Under the Accelerated Global Net Zero Scenario 12
1.3 Status of Net Zero Targets in Developing Asia 15
4.1 Climate Policies and Private Climate Capital 102
4.2 "Green" Monetary and Financial Policies of Global Central Banks, 2021 118
5.1 Main Features of National Emissions Trading Systems in Asia and the Pacific 144
5.2 Overview of Carbon Pricing Scenarios 157
5.3 Estimated Output Changes in Response to Carbon Pricing Throughout Asia and the Pacific 160
5.4 Estimated Changes in Exports in Response to Carbon Pricing Throughout Asia and the Pacific 161

Figures

1.1 Greenhouse Gas Emissions in Selected Regions and Subregions of the World, 1990–2021 6
1.2 Per Capita Emissions and Emissions Intensity in Developing Asia Versus Global Comparators 7
1.3 Emissions by Sector in Developing Asia, 2021 7
1.4 Share of Regional Emissions in Total Global Emissions for 2021, by Energy Subsectors 8
1.5 Share of Renewables in Total Installed Capacity in 2023, by Subregion 8
1.6 Renewable Energy Installation Costs and Generation 9
1.7 Public Preferences for Government Adaptation Measures and Funding Options 11
1.8 Possible Emissions Pathways for Developing Asia 13
1.9 Frequently Mentioned Mitigation Options in the Nationally Determined Contributions 14
of Developing Asia
1.10 Number of Nationally Determined Contributions That Mention Key Societal Issues 17
2.1 Annual Global Temperature Differences Compared to 1850–1900 Averages 22
2.2 Projected Changes in Days Above 35°C Annually Under a High End Emissions Climate Scenario 25
2.3 Projected Changes in the 6-Month Standardized Precipitation-Evapotranspiration Index 26
Under a High End Emissions Climate Scenario for Asia and the Pacific
2.4 Annual Damage from Sea-Level Rise Under a High End Emissions Climate Scenario 28
2.5 Annual Population Affected by Sea-Level Rise Under a High End Emissions Climate Scenario 29
2.6 Annual Additional Flood Damage Under a High End Emissions Climate Scenario 31
2.7 Annual Additional Population Affected by Flooding Under a High End Emissions Climate Scenario 32
2.8 Climate Change Effects on Crop Yields Under a High End Emissions Scenario as Shown 34
by Gridded Crop Growth Modeling
2.9 Cumulative Climate Change Impacts on Crop Yields Under a High End Emissions Scenario 35
2.10 Climate Change Impacts on Fishing Catch Under a High End Emissions Scenario 37
2.11 Climate Change Impacts on Forest Productivity Under a High End Emissions Scenario 38

2.12	Climate Change Impacts on Final Energy Demand Under a High End Emissions Scenario	40
2.13	Climate Change Impacts on Labor Productivity Under a High End Emissions Scenario	41
2.14	Framework to Quantify the Economic Effects of Climate Change Impacts	43
2.15	Loss of Gross Domestic Product due to Climate Change Under a High End Emissions Scenario	44
2.16	Sensitivity of Modeled Losses in 2070 due to Climate Change Under a High End Emissions Scenario	45
2.17	Composition of Losses in Gross Domestic Product in 2070 due to Climate Change Under a High End Emissions Scenario	46
2.18	Sector Composition of Modeled Losses due to Climate Change Under a High End Emissions Scenario	47
2.19	Economic Losses in 2100 due to Climate Change Under Various Climate Scenarios	48
3.1	Typology of Climate Change Adaptation Responses	58
3.2	Economic Framework for Adaptation Investments	59
3.3	Modeled Annual Adaptation Needs for Asia and the Pacific, 2023–2030	61
3.4	Global Annual Benefits Quantified from Flood Protection via Mangroves	63
3.5	Climate Change Adaptation Responses and Supporting Actions for Agriculture	70
3.6	Climate Adaptation Priorities for ADB Developing Member Countries	76
3.7	Key Adaptation Policy Instruments Across Asia and the Pacific	77
3.8	Stated and Modeled Annual Adaptation Finance Needs for 2023–2030, Total	79
3.9	Stated and Modeled Annual Adaptation Finance Needs for 2023–2030, per Capita	80
4.1	Public and Private Sector Shares of Climate Finance Across Asia and the Pacific, 2018–2019	93
4.2	Overview of Climate Finance Flows in Asia and the Pacific, 2018–2019	94
4.3	Drivers of Private Climate Investment	96
4.4	2030 Decarbonization Opportunities Under the International Energy Agency's Net Zero Emissions Scenario	97
4.5	Climate-Related Private Equity Investments by Sector	99
4.6	Sustainable Instrument Issuance in Emerging-Market and Developing Economies, by Instrument Type	100
4.7	Sustainable Debt Issuance in Emerging-Market and Developing Economies, by Region and Economy Type	100
4.8	An Overview of Green and Sustainable Taxonomies, 2023	101
4.9	Market Share of Environmental, Social, and Governance Funds, by Region	107
4.10	Consumption and Career Choices Made by Younger Generations, 2024	108
4.11	Climate Policy Uncertainty Index in the United States	110
4.12	Regulation and Standards Versus Subsidies in "Green" Industries	113
4.13	Examples of Holistic Policy Packages	114
4.14	Sustainable Finance Definitions	116
4.15	Sustainable Bonds to Support Transition Activities in Developing Asia	117
4.16	"Green" Policies and Regulations in Selected Developing Asian Economies	119
4.17	Share of the Top Three Economies in Total Production of Selected Minerals, 2022	121
5.1	Public Financial Support for Fossil Fuels in Developing Asia, 2015–2022	133
5.2	Fossil Fuel Subsidies in Developing Asia, 2022	134
5.3	Comparison of Carbon Prices in Selected Economies in Asia and the Pacific	146
5.4	Modeled Impacts of the Carbon Border Adjustment Mechanism and Changes to the European Union's Emissions Trading System	158
5.5	Modeled Impacts of the Carbon Border Adjustment Mechanism and Emissions Trading Across OECD Economies	159

Boxes

2.1   Impacts of Climatic Extremes on Crop Yields Under Climate Change   36
2.2   Impacts of Climatic Shocks on Physical and Human Capital in India   39
3.1   Reducing Vulnerability via Flood Shelters in Bangladesh   64
3.2   Challenges for Flood Forecasting and Early Warning Systems in Asia and the Pacific   66
3.3   Increasing Groundwater Availability by Managed Aquifer Recharge in Mongolia   67
3.4   Reducing Disease Through Climate-Resilient Municipal Services in Nepal   67
3.5   The Impacts of Emissions Mitigation on Agriculture   69
3.6   Digital Technologies to Contribute to Climate Adaptation   78
4.1   Climate Resilience Solutions in Asia and the Pacific   98
4.2   Fiji's First Sovereign Blue Bond   101
4.3   How the Inflation Reduction Act Increased Climate Finance in Asia and the Pacific   104
4.4   The European Union Sustainable Finance Disclosure Regulation   106
4.5   Climate Law and Sustainable Fund Holdings   111
5.1   Fossil Fuel Reform in India and Indonesia   136
5.2   Carbon Taxation Versus Emissions Trading System   142
5.3   Empirics of the European Union's Emissions Trading System   147
5.4   A Domestic Voluntary Carbon Market—The China Certified Emission Reduction Scheme   149
5.5   Concept of Carbon Market Linkage   150

FOREWORD

Asia and the Pacific is home to some of the world's most climate-vulnerable economies. At the same time, the region's strong growth, which has lifted millions out of poverty, contributes to more than half of global greenhouse gas emissions. As the climate crisis intensifies, this region finds itself at the center of both its worsening impacts and transformative potential solutions.

The *Asia–Pacific Climate Report* is a new knowledge initiative from the Asian Development Bank (ADB) to support Asia and the Pacific in combating climate change through policy reforms. This inaugural issue offers a comprehensive overview of the region's evolving climate landscape and examines various dimensions of the climate crisis, from the increasing frequency and severity of heat waves to the growing economic and social costs. The report underscores the urgent need for adaptation measures and emphasizes the importance of mobilizing resources to support the region's most vulnerable populations.

The report also examines the constraints facing governments in the region and proposes policy measures to address both adaptation and mitigation. For instance, the authors recommend integrating adaptation needs into development planning and budgeting processes across all sectors. On climate finance, the report discusses how governments can formulate consistent policies, promote information disclosure, and build climate-oriented financial systems to attract private capital. The report also highlights the significant space to improve policy instruments in the region and discusses carbon pricing—including carbon taxes—as a key policy for reducing carbon emissions.

ADB stands ready to collaborate closely with our developing member countries to operationalize these recommendations, including through financing to implement and evaluate policies. We will ensure that our resources and expertise are aligned to maximize development impact and achieve sustainable results.

The decisions made on climate action in the coming years will shape the region's future. We hope this publication inspires bold action and fosters regular discussion and collaboration to build a prosperous, inclusive, resilient, and sustainable Asia and the Pacific.

MASATSUGU ASAKAWA
President
Asian Development Bank

ACKNOWLEDGMENTS

It is our privilege to introduce the *Asia–Pacific Climate Report*. This new publication aims to shed light on emerging trends and address pressing challenges of climate action in the region. By providing comprehensive data and in-depth analysis, it seeks to strengthen climate policies and foster collaboration for more coordinated and effective mitigation and adaptation efforts.

This inaugural issue, entitled "Catalyzing Finance and Policy Solutions," provides an up-to-date overview of the climate landscape in Asia and the Pacific and discusses a variety of key topics, including public perceptions of climate change and policies, the impacts and costs of climate change under a high end emissions scenario, the urgent need for stronger adaptation responses, mobilizing private climate capital, and carbon pricing.

The report was authored by an Asian Development Bank (ADB) team consisting of staff from the Economic Research and Development Impact Department (ERDI) and the Climate Change and Sustainable Development Department (CCSD), and external consultants: Yi Jiang (lead), Madhavi Pundit (lead), David Raitzer (lead), Shu Tian (lead), Sabah Abdullah, Sanchita Basu-Das, Francesco Bosello (consultant), Lorenza Campagnolo (consultant), Brendan A. Coleman, Virender Kumar Duggal, Neil Foster-McGregor, Alessio Giardino, Bianca Gutierrez, Kate Hughes, Jules Hugot, Johannes Hunink (consultant), Esmyra Javier, Jong Woo Kang, Kijin Kim, Xianfu Lu (consultant), Ma. Noelle Manahan (consultant), Arturo M. Martinez Jr., Gabriele Mansi (consultant), Michael Mehling (consultant), Homer Pagkalinawan, Martino Pelli, Manisha Pradhananga, Alexander Raabe, Aisha Reynolds (consultant), Arghya Sinha Roy, Melanie Grace A. Quintos, Eelco van Beek (consultant), Maria van Veldhuizen (consultant, Green Ink), Mai Lin Villaruel, Michalis Vousdoukas (consultant), Paul Watkiss (consultant), and Junjie Zhang (consultant).

Guidance and support were provided by the ERDI and CCSD management: Deputy Chief Economist Joseph E. Zveglich Jr., Deputy Director General Chia-Hsin Hu, Senior Director Toru Kubo, Director Abdul Abiad, Director Lei Lei Song, Director Noelle O'Brien, and ourselves. Abdul Abiad, Yi Jiang, and Declan Magee were responsible for the overall coordination of the report preparation. The Expert Advisory Committee for the report consisting of Rachel Kyte, Ji Zou, and Warren Evans, also reviewed the report and provided valuable comments that helped refine the report. Comments and suggestions from the Asia–Pacific Climate Report 2024 Background Papers' Workshop held on 8–9 May 2024, the bank-wide seminar on 1 October 2024, and ADB's internal review improved the earlier drafts. The report also benefited from comments and recommendations of the ADB Office of the President and the Board of Directors.

The report draws on background papers and materials prepared by Sabah Abdullah, Abdul Abiad, Jose Ramon Albert, Dorothy Bantasan, Christopher Beaton, Laura Cameron, Nicolas Charette, Gabriele Ciminelli, Sanchita Basu-Das, Nhat Do, Joeffrey Drouard, Jackson Ewing, Neil Foster-McGregor, Joseph Francois, Yazhen Gong, Shreekant Gupta, Shweta Gupta, Mengfu Han, Karthik Iyer, Daniel te Kaat, Tara Laan, Anil Markandya, Arturo M. Martinez Jr., Michael Mehling, Joseph Mik, Jason Mortimer, Papa Orgen, Martino Pelli, Madhavi Pundit, Alexander Raabe, Swasti Raizada, David Raitzer, Deepak Sharma, Daniel Marcel te Kaat, Milan Thomas, Shu Tian, Yuanjie Tian, Jeanne Tschopp, Prashant Vaze, Olivier Bois von Kursk, Roberton Williams, Eugene Wong, Fei Xie, and Junjie Zhang.

Paul Dent edited the entire report, while Tuesday Soriano copyedited and proofread it. Dorothy Bantasan, Dyann Buenazedacruz, Bhabhani Dewan, Aileen Gatson, Eugene Creely Ingking, Jade Laranjo, Jindra Nuella Samson, Iva Sebastian-Samaniego assisted in proofreading. Margaret Ruth Griffiths, Ma. Josefa Quitazol Gonzalez, Cherry Lynn Zafaralla, and Eric van Zant edited the background papers. Hannah Althea Estipona formatted the background papers. Joe Mark Ganaban typeset and laid out the report and created the cover design and header graphics. Dorothy Bantasan, Aileen Gatson, Lotis Quiao, and Jasmin Sibal provided overall administrative and logistical support. The Department of Communications and Knowledge Management, led by Simon Pollock and Terje Langeland, coordinated report dissemination, with support from Lean Alfred B. Santos, Duncan McLeod, Anna Sherwood, Keech Hidalgo, and Maricris Adan. Ralph Romero designed the landing webpage.

ALBERT F. PARK
Chief Economist and Director General
Economic Research and Development
   Impact Department
Asian Development Bank

BRUNO CARRASCO
Director General
Climate Change and Sustainable
   Development Department
Asian Development Bank

DEFINITIONS

The economies discussed in this report are classified by major analytic or geographic grouping.
For the purposes of this publication, the following apply, unless otherwise specified:

- "Developing Asia" and "Asia and the Pacific" are used interchangeably in this report to refer to the 46 regional members of the Asian Development Bank listed by geographic grouping as follows:

  - Caucasus and Central Asia comprises Armenia, Azerbaijan, Georgia, Kazakhstan, the Kyrgyz Republic, Tajikistan, Turkmenistan, and Uzbekistan.

  - East Asia comprises the People's Republic of China; Hong Kong, China; Mongolia; the Republic of Korea; and Taipei,China.

  - South Asia comprises Afghanistan,[a] Bangladesh, Bhutan, India, Maldives, Nepal, Pakistan, and Sri Lanka.

  - Southeast Asia comprises Brunei Darussalam, Cambodia, Indonesia, the Lao People's Democratic Republic, Malaysia, Myanmar, the Philippines, Singapore, Thailand, Timor-Leste, and Viet Nam.

  - The Pacific comprises the Cook Islands, Fiji, Kiribati, the Marshall Islands, the Federated States of Micronesia, Nauru, Niue, Palau, Papua New Guinea, Samoa, Solomon Islands, Tonga, Tuvalu, and Vanuatu.

Unless otherwise specified, the symbol "$" and the word "dollar" refer to United States dollars.

---

[a]   ADB placed its regular assistance to Afghanistan on hold effective 15 August 2021.

ABBREVIATIONS

ADB	Asian Development Bank
CBAM	Carbon Border Adjustment Mechanism
CDM	Clean Development Mechanism
CO_2	carbon dioxide
COP	Conference of the Parties
CPI	Climate Policy Initiative
PRC	People's Republic of China
DIVA	Dynamic Interactive Vulnerability Assessment
ESG	environmental, social, and governance
ETS	emissions trading system
EU	European Union
GCM	global circulation model
GDP	gross domestic product
GHG	greenhouse gas
GLOFRIS	Global Flood Risk with Image Scenarios
GSS	green, social, and sustainability
$GtCO_2e$	gigaton of carbon dioxide equivalent
ICAO	International Civil Aviation Organization
ICES	Intertemporal Computable Equilibrium System
ICVCM	Integrity Council for the Voluntary Carbon Market
IPCC	Intergovernmental Panel on Climate Change
ISIMIP	Inter-Sectoral Impact Model Intercomparison Project
ISSB	International Sustainability Standards Board
ITMO	Internationally Transferred Mitigation Outcome
JCM	joint crediting mechanism
LDE	least-developed economy
MDB	multilateral development bank
MRV	monitoring, reporting, and verification
$MtCO_2e$	metric ton of carbon dioxide equivalent
NAP	National Adaptation Plan
NBS	nature-based solution
NCCC	National Communications on Climate Change
NDC	nationally determined contribution
OECD	Organisation for Economic Co-operation and Development
PACM	Paris Agreement Crediting Mechanism
PPP	purchasing power parity
RCP	Representative Concentration Pathway
ROK	Republic of Korea

SPEI	Standardized Precipitation-Evapotranspiration Index
SOE	state-owned enterprise
SSP	Shared Socioeconomic Pathway
UN	United Nations
UNFCCC	United Nations Framework Convention on Climate Change
VCM	voluntary carbon market
VCMI	Voluntary Carbon Markets Integrity Initiative
WASH	water, sanitation, and hygiene

HIGHLIGHTS

Evolving climate risks in Asia and the Pacific*

- **Continuation of implemented policies will lead to dangerous levels of global warming.** With the policies currently being implemented, around 3°C of warming is expected within this century. While pledges in nationally determined contributions (NDCs) could reduce this to 2.4°C, taking into account feedback effects implies that warming could happen faster than conventionally estimated.

- **Developing Asia both suffers from and contributes to the climate crisis.** The emissions intensity of the region's economies has decreased by over 50% since 2000. However, the region still generated about half of global greenhouse gas emissions in 2021, driven by rising domestic consumption, energy demand, and production. Without stronger adaptation and mitigation efforts, the region faces severe impacts and substantial economic losses.

- **Many impacts in the Asia and Pacific region will be greater than indicated by global trends.** Under a high end emissions scenario, relative sea-level rise and temperature increases in the region will exceed global averages. The destructive power of typhoons and cyclones will double, and flood losses will increase due to more concentrated rainfall and glacial melt. Climate-dependent sectors such as agriculture, forestry, and fisheries will face diminished output. Hotter temperatures will reduce labor productivity, erode human and social capital, and increase mortality and health risks.

- **Climate change reduces gross domestic product in the region by 17% in 2070 under a high end emissions scenario.** Sea-level rise and labor productivity effects dominate the losses, with relative effects concentrated in lower-income and fragile economies and those with large coastal populations. The impacts are regressive and hit vulnerable and poor people the hardest.

- **Public concern about climate change is high, as is support for ambitious climate action.** The ADB climate change perception survey conducted in 2024 finds that 91% of respondents in 14 economies view climate change as a serious problem. Improving policy design, addressing inequalities, and increasing awareness about climate action can boost public support.

Accelerating climate change adaptation

- **Adaptation is key to addressing large impacts.** In the near term, only adaptation can alter climate losses. Effective adaptation policy must address both slow onset changes and a greater frequency and/or intensity of fast onset extreme events, while at the same time considering other development needs.

---

* This document summarizes the key messages of the *Asia–Pacific Climate Report 2024: Catalyzing Finance and Policy Solutions*. The report focuses on critical climate change issues for the Asia and Pacific region that have not been recently covered in detail.

- **Early and coordinated investments are needed to avoid damage and enable transformational responses.** Realizing adaptation opportunities requires enhanced analysis of climate change impacts and adaptation needs upstream in development planning. Although most countries have prepared National Adaptation Plans (NAPs) and included adaptation in their NDCs, they can still better integrate adaptation into medium-term development plans and expenditure frameworks.

- **Market reforms can enhance adaptive capacity by correcting market distortions.** Adaptation is facilitated when climate signals are well transmitted and market actors have the freedom to respond. Strengthening property rights can encourage better incorporation of climate risks. Removing subsidies for water and other natural resources can curb the overuse that exacerbates scarcity and damage. More open trade can diversify supply chains, stabilize markets, and promote the spread of adaptation technologies.

- **Both public and private sources are key to closing the adaptation financing gap.** Annual financing needs for adaptation in the region are estimated to be between $102 billion and $431 billion. Nearly half of modeled needs are for coastal and river flood protection. The needs far exceed the approximately $34 billion of tracked adaptation finance that was committed in the region in 2021–2022.

- **Continuous reassessment of climate change impacts and adaptation actions is essential.** Policymakers must continually revisit and improve predictions of impacts and systematically test adaptation measures to ensure that adaptation policy addresses evolving challenges. The results of this ongoing reassessment must be communicated continuously to ensure that climate risks are internalized in planning and investment processes.

Scaling up private climate capital

- **Private capital is increasingly becoming a source of climate finance, but more needs to be done.** Climate-related risks and regulatory changes are pushing more private capital into climate-aligned investments. However, policy uncertainties and misalignments, unreliable information disclosure, and weak enabling financial markets are constraints to raising private climate capital.

- **Greater policy certainty and financial sector reform can attract private climate capital.** Governments can adopt a comprehensive set of aligned policies featuring a mix of price and nonprice policies, such as carbon pricing and clean energy subsidies, to enhance the profitability of climate investments and provide consistent incentives. Policymakers can develop climate-oriented financial systems through the adoption of sustainability disclosure standards to reduce transaction costs and compliance risks for investors.

- **The public sector can help overcome barriers to private climate investment.** Solutions like blended finance can facilitate the use of de-risking instruments, such as first-loss guarantees, insurance, and credit enhancements, to attract private investors. Blended finance can also generate demonstration effects for new technologies. The public sector can strengthen the investment environment by liberalizing markets, reducing entry barriers, and overcoming market rigidities such as long-term power purchase agreements. Multilateral development banks can support these efforts with their capital and their menu of instruments.

Carbon pricing for more cost-effective mitigation

- **Carbon pricing instruments are gaining momentum in the region.** Eight national initiatives are currently being implemented, while three countries are considering a domestic emissions trading system (ETS). However, these efforts are challenged by the continued significant financial support for fossil fuel subsidies.

- **Carbon taxation can be a powerful mechanism to reduce emissions cost-effectively.** A carbon tax is not only an efficient approach to reducing emissions, but can also raise significant revenues that could be redirected to the highest-value public uses. Existing studies suggest that the adverse impact of a carbon tax on growth, inflation, and overall employment is likely to be small. Furthermore, policymakers have options to make the distributional effects more progressive, including through targeted transfers, anti-poverty programs, or reducing regressive taxes.

- **There are several important lessons on ETS for the region.** Efficiency-based targets balance emission reduction and production expansion goals but may not reduce emissions sufficiently. Institutional and regulatory frameworks as well as monitoring, reporting, and verification systems are critical and require significant effort and careful legislation. The ETS can work more effectively with more carbon prices passed through the power market. Allowing offsets in an ETS can increase flexibility to help meet targets.

- **Developing economies can leverage international carbon markets to meet climate goals.** Purchasing credits can help meet NDCs, especially when domestic abatement costs are high. However, developing national carbon market strategies is essential for effective engagement with international markets, effective NDC accounting frameworks, and voluntary market participation.

EXECUTIVE SUMMARY

This report consists of five chapters that focus on critical climate change issues for the Asia and Pacific region that have not been recently covered in detail. Chapter 1 provides an overview of the regional context for climate action. Chapter 2 examines the potential economic impacts of a high end emissions scenario. Chapter 3 identifies priority adaptation actions to help address the impacts of climate change. Chapter 4 considers ways to better mobilize private sector finance for mitigation and adaptation. The fifth and final chapter discusses how carbon pricing can be designed and implemented more effectively to achieve successful mitigation.

1. Addressing the Climate Crisis in Asia and the Pacific

- **Developing Asia both suffers from and contributes to the climate crisis.** Global greenhouse gas (GHG) emissions continue to rise and without stronger mitigation efforts, the region will experience more extreme heat, floods, storms, and infectious diseases. Just 9 years after the adoption of the Paris Agreement on Climate Change, inaction has put the 1.5°C ambition nearly out of reach. Exceeding this target could significantly increase the risk of overshooting tipping points, leading to even more harmful impacts and related losses and damages. Nonetheless, the global commitment to climate adaptation and mitigation offers hope. Accelerating the transition to net zero emissions and scaling up investments in advanced climate technologies and nature-based solutions can foster low-carbon, climate-resilient economic growth.

- **The region generated about half of global GHG emissions in 2021, surpassing many historically high emitters.** The increase over the past 2 decades has been driven by rising domestic consumption, energy demand, and production for advanced economies. The People's Republic of China (PRC) accounted for two-thirds of the increase, while South Asia contributed 19.3% and Southeast Asia 15.4%. The energy sector is the largest contributor to the region's emissions (77.6%), primarily due to a heavy reliance on fossil fuels. Other high-emitting sources are agriculture (10.6%) and industrial processes (8.6%).

- **Developing Asia has taken significant steps for decarbonization, but more ambitious mitigation action is necessary to slow global climate change.** Since 2000, the emissions intensity of the region's economies has decreased by over 50%, although it remains above the global average. Further, the region has invested more in renewable energy than any other world region. An accelerated transition to net zero would require substantially more investment, but credible pathways to achieve this have been identified. The costs of inaction would be far greater, while rapid decarbonization brings significant co-benefits, including improved public health, fewer premature deaths, reduced agricultural losses, and the creation of high-quality jobs in emerging sectors.

- **The region is well positioned to benefit from the global transition to net zero.** With substantial renewable energy generation potential, the region can produce some of the world's cheapest renewable electricity, helping to reduce carbon intensity in industry and energy sectors. Its fast-growing and innovative economies, large workforce, high-quality technical education, and strong manufacturing base provide an opportunity to develop and supply the technologies needed for global decarbonization. Moreover, the growth of renewable energy manufacturing can enhance energy security and contribute to sustainable development in the region.

- **Developing Asia is committed to tackling climate change, but there is scope to raise ambitions.** Forty-four economies, representing 98% of the region's emissions, have ratified the 2015 Paris Agreement and submitted nationally determined contributions (NDCs) with commitments to cut GHG emissions and adapt to climate change. In addition, 36 of these economies have announced or adopted net zero targets. However, current actions fall short of what is needed to limit the global temperature rise to 2°C or, ideally, 1.5°C. Both the NDCs and net zero targets lack concrete, sector-specific emission reduction road maps. Key challenges of implementing decarbonization plans include institutional capacity gaps, governance issues, insufficient financing, limited access to technology and expertise, and vulnerability to climate events that divert resources from decarbonization efforts. More evident support is needed to bridge the gaps between the required ambition and commitments, commitments and concrete road maps, and road maps and implementation.

- **Public concern about climate change is strong, along with support for ambitious climate action.** The ADB climate change perception survey conducted in 2024 finds that 91% of people in 14 economies in developing Asia view climate change as a serious problem. Additionally, 84% of respondents report that they and their families are already experiencing climate change impacts or expect to within the next decade. A majority support measures limiting emissions, investment in low-carbon and resilient infrastructure, and the introduction of a carbon tax. While there is evident support for climate measures in developing Asia, concerns about their effectiveness and costs remain. Enhancing awareness about climate actions while ensuring that policies are well-designed and address existing inequalities can alleviate doubt and boost public support.

2.   Impacts and Costs of Climate Inaction for Asia and the Pacific

- **Implemented policies will lead to dangerous levels of global warming.** The 10 warmest years since global temperatures started being recorded in 1850 all occurred in 2014–2023, and 2023 contained the warmest months on record. If only the policies that have already been implemented are continued, around 3°C of mean warming can be expected, while the national mitigation pledges in the NDCs that depend on implementing new policies could reduce warming to a mean projected level of 2.4°C. Emissions trends up until 2021 (the most recent year of comprehensive data) show no substantial deviation from scenarios with much more warming than these values.

- **Feedback effects mean that climate change could happen faster than anticipated.** Emissions of methane, a potent GHG from natural ecosystems are rapidly rising, due to increases in temperature and rainfall. Antarctic sea temperatures have been at record highs in 2024, even as sea ice is at an all-time low, reducing the reflectance of solar radiation. Fires are increasingly turning forests into carbon sources rather than carbon sinks. As the models used to determine the warming expected under "implemented policies" do not adequately capture these interactions, there could be more warming than conventionally estimated, and the types of changes typifying a "high end emissions" scenario (termed Shared Socioeconomic Pathway 5-8.5 by the Intergovernmental Panel on Climate Change) could still occur.

- **Changes in the Asia and Pacific region will be greater than global trends indicate.** Temperature rises over inhabited land areas will exceed global averages, and increases will be concentrated in certain seasons and subregions. Heat waves will dramatically increase under a high end emissions scenario, with average days above 35°C increasing by several times by 2100. Rainfall will become more concentrated, with peak daily precipitation increasing by 30%–50% by 2100, even as droughts become more frequent. At the same time, the destructive power of typhoons and cyclones will approximately double in the region.

- **Sea-level rise will increasingly put coastal assets and populations at risk.** Rates of relative sea-level rise in key areas of the Asia and Pacific region are about double the global average, and the region contains 70% of global population threatened by sea-level rise. Around 300 million people in the region could be threatened by coastal inundation in the event of sea ice instability. This will be exacerbated by increasingly destructive storm surges. New modeling suggests that trillions of dollars of coastal assets in the region could be damaged annually by 2070 under a high end emissions scenario, while tens of millions of people could be affected each year.

- **Flood losses will dramatically rise under climate change.** With more rainfall and glacial melt, flows in major river systems will increase, especially in South Asia. In addition, more concentrated and intense rainfall, as well as more storm events, will lead to increased inundation in low-lying areas and landslides in mountainous areas. Models suggest that around $1 trillion of annual flood damage could occur by 2070 under a high end emissions scenario, primarily in South Asia.

- **Climate-dependent sectors, including agriculture, forestry, and fisheries, will face diminished output.** Although higher carbon dioxide (CO_2) concentrations may benefit crop growth by speeding up photosynthesis, increasing water stress will increasingly impact yields. New empirical research suggests that wheat yields could be reduced by up to 45% and maize yields by over 20% in parts of developing Asia under a high end emissions scenario by 2070. Forests have a species composition that cannot change quickly in response to climate shifts, and they face increasing risk of pest and disease attacks, dieback, and fire under climate change. Warming of seawater and acidification by increasing CO_2 concentrations put reefs and fisheries at substantial risk. Leading models show that both forestry and fisheries yields will substantially decline under a high end emissions scenario.

- **Hotter temperatures will reduce labor productivity as well as human and social capital.** With longer and more frequent heat waves, much of developing Asia and the Pacific will face increasing periods during which temperatures constrain labor supply and labor productivity. Empirical estimates suggest that a high end emissions scenario would uniquely reduce labor productivity in the region, with up to a 30% reduction by 2070. Extended heat waves may also lead to higher workplace accident rates, increased conflict, and lower learning rates in schools. More heat will increase cardiovascular mortality and exacerbate more extreme event injuries and broader transmission of vector-borne illnesses under climate change. Many of the adverse effects on human and social capital will disproportionately affect girls and women.

- **Climate change can reduce gross domestic product (GDP) in Asia and the Pacific by 17% by 2070 under a high end emissions scenario.** Considering shocks to agriculture, forestry, fisheries, and other sectors; river flooding; sea-level rise; energy; and labor productivity in an economy-wide model reveals high economic vulnerability. Sea-level rise and labor productivity effects dominate losses, with relative effects concentrated in lower-income and fragile economies such as Bangladesh and the Pacific island countries, or those with extensive coastal populations such as Viet Nam. If more complex interaction effects are considered, the losses could be much higher. Extrapolation of patterns of loss relative to warming suggests that losses could reach 41% by 2100 under a high end emissions scenario.

- **Impacts will hit vulnerable and poor people the hardest.** With economic losses highest for some of the lowest-income economies, impacts are regressive. The impact channels considered also imply that those with the lowest incomes will suffer most. Poor people are more likely to live in areas exposed to flooding and storm surges, depend on natural resources sectors like agriculture for their livelihoods, work in sectors exposed to increasing heat, and spend much of their income on food.

3.   Accelerating Climate Change Adaptation in Asia and the Pacific

- **Adaptation is key to addressing large impacts locked in from emissions to date.** There are long lags in the earth's climate system, so reductions in emissions will only gradually change the pace of warming. In the near term, only adaptation can alter climate losses, while in the long term, adaptation and mitigation must also address losses that are likely to be much more profound given implemented policies and emissions trajectories. Effective adaptation policy must address both slow onset changes and a greater frequency and/or intensity of fast-onset extreme events, while at the same time considering other development needs. Mechanisms are also needed to address residual loss and damage that remain after adaptation measures.

- **Early and coordinated investments are needed to avoid damage and enable transformational responses as changes become more severe.** There are extensive "no regret" adaptation opportunities that are justified by large economic benefits, high-return opportunities to incorporate climate-smart design into new investments, and low-cost preparatory actions to enable more resilience options in the future. Realizing these opportunities requires enhanced analysis of climate change impacts and adaptation needs upstream in development planning processes, rather than in the current fragmented approach that characterizes efforts in many countries. Climate change introduces uncertainty, so adaptation requires a dynamic process of readjustment as understanding grows and evolves.

- **A wide range of investments are needed in vulnerable sectors and systems.** Adaptation is needed in nearly every sector, although needs vary among sectors and systems. Models find that $102 billion is needed per year in adaptation investment in the Asia and Pacific region up to 2030. Coastal protection and river flood protection account for nearly half of this, with additional needs for agriculture, fisheries, health, social protection, and other infrastructure. Nearly every sector has both hard infrastructure needs to preserve functionality and serve new needs in the face of climatic changes, and soft needs to improve planning, coordination, flexibility and governance. Deploying systems that can cope with the volatility and unpredictability of future weather and deploy more knowledge-intensive solutions will require enhanced technical and institutional capacity in many areas.

- **Market reforms can help to enable adaptive capacity.** Individuals and firms have a natural incentive to pursue climate change adaptation if markets are functioning well, as benefits are local and appropriable. Climate change adaptation is facilitated when climate signals are well transmitted and market actors have freedom to respond. In Asia and the Pacific, key markets, such as for land, are subject to incomplete property rights and other distortions, which restrict incorporation of climate risks. Subsidies for water and other natural resources encourage overuse that exacerbates scarcity and damage. International trade can be an important means of adapting to climate change, but many countries in the region are imposing trade barriers. More open trade can help build resilience to climate-related shocks by diversifying supply chains, stabilizing markets, and spreading adaptation technologies.

- **All economies in the region have adaptation policies, but many could be strengthened.** Most have prepared National Adaptation Plans (NAPs), and most NDCs from the region also include priorities on adaptation. While many of these documents cost climate adaptation measures, there remains scope to use economic analysis to help set priorities. More broadly, there remains a need to integrate adaptation into medium-term development plans and expenditure frameworks.

- **Adaptation investment needs are far in excess of financing flows.** A review of financing needs identified by countries in the region indicates annual targeted investment of $431 billion. Although, in aggregate, this is higher than modeled needs of $102 billion, what is clear is that identified needs are far in excess of the approximately $34 billion of tracked adaptation finance that was committed in the region in 2021–2022, or the $11 billion that the region received from international sources. Only around 1% of known regional adaptation finance came from the private sector.

- **To fill the financing gap, adaptation finance should be mobilized from both public and private sources.** Adaptation investment plans can be an important tool to engage stakeholders to fill the gap between NAPs and bankable adaptation investments, at the strategic rather than project level. They can provide more coherent consideration of economic analysis (including positive externalities) and financial assessments, and more focus on using public, private, and blended solutions. Catalytic instruments, including innovative risk-sharing and guarantees, can be used to blend private and public finance to lower initial and life-cycle costs for adaptation infrastructure.

- **Continuous reassessment and communication of potential climate change impacts and adaptation actions are warranted, given uncertainties.** Climate change adaptation policies are set in the context of (i) uncertainty about future GHG emissions, (ii) climate model uncertainty about the climate and weather changes that result from emissions, (iii) sector model uncertainty about the biophysical impacts of climate and weather changes, and (iv) socioeconomic uncertainty about how development will condition people's exposure and vulnerability. This means that there is a need for policymakers to continually revisit and improve predictions with new science, including via updated climate models, downscaling models, sector models, and economic models. Similarly, many adaptation measures involve innovations and new approaches that benefit from systematic testing. To ensure that climate risks are internalized in all relevant planning and investment processes, they need to be continuously communicated.

4. Scaling Up Private Climate Capital for Mitigation and Adaptation

- **Private sector capital is essential to meet vast climate finance needs.** While climate finance has been growing globally, it needs to keep rising as a large gap remains between needs and available financing. Private capital is the key to closing this funding gap. During 2021–2022, the private sector provided 49% of global climate finance, although this was only 40.2% in developing Asia. Given limited public resources and competing policy priorities, the private sector's share is expected to rise to around 90% by 2030.

- **Emerging business opportunities boosted by climate policies and financial market development facilitate the scale-up of private climate capital.** Both climate mitigation and adaptation offer major investment opportunities, with commercially viable opportunities in renewable energy production, upgrading electricity grids, electric vehicle production and charging infrastructure, and water-saving technologies. Climate policies such as carbon pricing and clean energy subsidies help enhance the profitability and commercial viability of climate investments. Financial market development, including green and sustainable bonds, sustainability-linked bonds, and transition bonds help direct private capital into climate investments. Climate investments also contribute to nonpecuniary gains for investors, such as improved reputation and stakeholder partnerships.

- **Greater climate-related risks and the associated shifts in the regulatory landscape are pushing significantly more private capital into climate-aligned investments.** Climate-related risks threaten business viability due to supply disruptions, asset damages, and demand shocks, leading firms to invest in climate-resilient technologies and adapt their business processes. Institutional investors are shifting toward climate-aligned assets to reduce their exposure to climate-related risks, adjust to the changing preferences of stakeholders, and comply with emission reduction targets, including those deriving from carbon-border adjustment mechanisms. Climate-related disclosure helps markets differentiate firms' climate-related risk exposure and progress made in addressing them, thus benefiting the more sustainable firms.

- **Uncertainty and misaligned climate policies weaken the incentives for private climate investment.** Various political and economic considerations can weaken the implementation of climate policies, causing policy uncertainty and hampering confidence in climate investments. Our analysis shows that investors hold more sustainable assets when countries adopt climate laws, a signal of reduced climate policy certainty. Misaligned policies reduce incentives for private investors. For example, incentives to invest in low-carbon projects weaken when carbon pricing, green subsidies, and fossil fuel subsidies exist at the same time.

- **Increased policy certainty via comprehensive and aligned policies can generate consistent incentives to attract private climate capital.** Governments may adopt a comprehensive set of climate policies featuring a mix of price and non-price policies to enhance profitability of climate investments and incentivize private capital. Governments must ensure that policies are aligned with each other to offer consistent incentives for climate-aligned investments. Holistic policy packages can help mobilize private climate capital by pairing climate policies with supporting infrastructure and an enabling ecosystem. One example is combining transportation emissions efficiency standards with the establishment of electric vehicle charging stations.

- **A lack of reliable information disclosure and enabling financial markets is a constraint on raising private climate capital.** Limited climate risk data, risk management tools, and climate disclosures constrain climate-oriented private investments. Sustainability disclosure is largely practiced by publicly listed companies, while many nonpublicly listed companies, especially micro, small, and medium-sized enterprises, lack the monitoring capacity and resources to participate. Sustainable bonds and loans remain a small fraction of overall corporate financing in developing Asia. The absence of an agreed-upon transition taxonomy and credible transition pathways also constrain private investments. Multiple overlapping taxonomies raise transaction costs and may even disrupt financial intermediation for international climate finance. Underdeveloped capital markets in some developing Asian economies limit their capacity to mobilize private climate capital.

- **Developing a climate-oriented financial system and accelerating sustainability disclosure can further unlock private climate capital.** Building a climate-oriented financial system helps promote private climate capital by mainstreaming the consideration of climate-related risks that undermine investment decisions. Policymakers can accelerate the adoption of globally compatible sustainability disclosure standards, such as those of the International Sustainability Standards Board, to effectively reduce the transaction costs and compliance risks of investors. Governments need to equip small and medium-sized enterprises with capacity and resources for sustainability disclosure. Regulators need to lead the development of transition taxonomies and support institutions to develop credible transition pathways to raise more transition finance. Regulators must develop globally compatible green and transition taxonomies to raise climate finance from a global private finance pool.

- **Several challenges weaken the commercial viability of climate-oriented projects.** Many climate-oriented projects face challenges such as lack of insurance, high up-front costs, and relatively long payback horizons. For example, renewable energy faces geopolitical risks due to concentrated supplies of critical minerals. The commercial viability of some climate investments is weakened by high preparation costs and socio-environmental costs. Foreign private investors also face higher counterparty credit risk, currency risk, regulatory risk, and repatriation risk, especially when investing in emerging economies. This further reduces commercial viability.

- **The public sector can help overcome barriers to private climate investment by de-risking investments and creating an enabling investment environment.** Governments and development institutions can stimulate private investments through de-risking instruments such as first-loss guarantees, insurance, and credit enhancements. Solutions like blended finance can enable adoption of innovative structures and de-risking instruments to attract private investors with different risk-return preferences and mandates. Blended finance also can generate demonstration effects for new technologies. The public sector can strengthen the investment environment to enable private sector participation by liberalizing markets, reducing entry barriers to help scale new technologies, and overcoming market rigidities such as long-term power purchase agreements.

- **Multilateral organizations and global initiatives can help stimulate private financing in developing Asia.** Multilateral organizations and global or regional initiatives help boost private finance through schemes such as the Multilateral Investment Guarantee Agency and investments such as policy-based lending or project support facilities. Technical assistance can help specify how climate commitments can include necessary policy reform and implementation. Multilateral development banks can add value by helping governments develop an enabling market environment, strengthen climate investment pipelines via better project preparation that engages the private sector and establish modalities and partnerships with global initiatives and multilateral organizations to improve governments' capacity for policy and market development.

5. Carbon Pricing as a Key Policy for Climate Change Mitigation

- **There is growing momentum for applying carbon pricing instruments in Asia and the Pacific.** As an integral element of the broader climate policy architecture, carbon pricing plays a critical role in enabling developing countries in the region to achieve climate mitigation targets and green economic growth. Eight national initiatives have been implemented or are being developed in the region, including carbon taxes in Japan and Singapore and emissions trading systems (ETSs) in Australia, the People's Republic of China (PRC), Indonesia, Kazakhstan, the Republic of Korea, and New Zealand. Additionally, the Philippines, Thailand, and Viet Nam are considering the adoption of a domestic ETS. Momentum is also growing in the region for economies to participate in international carbon markets under Article 6 of the Paris Agreement.

- **Developing Asia provided $600 billion in public financial support for fossil fuels in 2022, which acts as a negative carbon price and reinforces dependence on fossil fuels.** The support included $440 billion in subsidies, at least $146 billion in capital investment by state-owned enterprises and $14 billion in international public finance. Most of the subsidies reduce prices of fossil fuel, leading consumers to use more fossil fuels than they otherwise would. Financial support for new production capacity increases fossil fuel investment and therefore locks countries into fossil supply chains, creating the risk of stranded assets. Fossil fuel subsidies are a costly and counterproductive mechanism to support energy access if they are not targeted and complemented with measures that incubate clean energy alternatives. Rationalizing and phasing out fossil fuel subsidies has been slow, despite pledges made by governments.

- **Carbon taxation can be a powerful mechanism for cost-effectively reducing GHG emissions.** By letting polluters decide how to reduce their emissions, a carbon tax ensures that the methods chosen minimize costs. A carbon tax can also raise significant revenue more efficiently because a well-designed carbon tax is harder to avoid or evade than other commonly used taxes. This is a particular advantage in economies with substantial informal sectors. An ideal carbon tax should reflect the harm caused by carbon emissions, cover a broad range of GHG emissions, and be collected "upstream." The revenue from carbon taxation should be redirected to the highest-value public uses.

- **A carbon tax is expected to have moderate macroeconomic effects while not being regressive.** Although further evidence is needed, existing studies suggest that the adverse impacts of a carbon tax on both growth and inflation are likely to be small. The employment effects of a carbon tax are primarily a shift in jobs from carbon-intensive to lower-emission sectors, with minimal aggregate effects on employment. Policies such as preannouncement, phased implementation, and worker retraining can facilitate smoother transitions for affected workers. Policymakers can make the distributional effects more progressive by using tax revenues for targeted transfers, anti-poverty programs, or reducing regressive taxes.

- **Experience in designing and operating ETSs in the region holds several lessons for economies looking to establish one.** Efficiency-based targets balance the goals of reducing emissions and expanding production, but they may limit the ability of the ETS to significantly curb absolute emissions. To mitigate this, countries should incorporate clear timelines and strategies for ensuring ETS emission impacts and transitioning to pure cap-and-trade systems. Second, institutional and regulatory frameworks are pivotal for the smooth operation of an ETS. When developing an ETS, Asian economies should dedicate significant work and resources to develop and refine their monitoring, reporting, and verification systems. Legislation should mandate facilities that have an impact report their annual GHG emissions data. In addition, the heavily regulated power sectors will hinder carbon price pass-through in many developing Asian economies, muting the overall ETS. ETS architects and governments should coordinate closely with power market reformers to enhance the effectiveness of ETS pricing mechanisms. Finally, allowing for offset use in an ETS can add flexibility for regulated entities to meet emissions reduction targets.

- **Developing economies in Asia and the Pacific can leverage international carbon markets to meet domestic climate goals.** While many focus on exporting carbon credits, there is also potential for purchasing credits to meet NDCs, especially when domestic abatement costs are high. Private sector involvement can help overcome the political challenges of using public funds for carbon credit purchases. Although the region has extensive experience with carbon market mechanisms, countries face challenges such as lack of national strategies, governance, institutional capacity, and carbon market expertise. The development of holistic national carbon market strategies is essential for engaging effectively with international markets. These strategies should strike a balance between selling mitigation outcomes and meeting NDC targets and promote the integration of compliance and voluntary markets. Additionally, effective NDC accounting frameworks are crucial for authorizing and tracking international transfers of mitigation outcomes. Governments must also promote and potentially regulate voluntary market activities to enhance market participation.

- **The carbon pricing impact on competitiveness between countries and regions can be mitigated with proper policy responses.** The effects will depend on both current competitiveness and carbon intensities in affected industries. Potential responses include output-based rebating and/or border carbon adjustments. Output-based rebating provides tax rebates proportional to production in emitting industries, while border adjustments tax carbon-intensive imports. Border carbon adjustments offer better incentives, enforce a level playing field in trade, and help reduce carbon leakage. Yet, they cannot avoid distortions and leakage unless carbon pricing schemes are extended to a broader set of countries worldwide.

Chapter 1

Addressing the Climate Crisis in Asia and the Pacific

1.1 The Need for Accelerated Climate Action

Climate change is a defining challenge of the 21st century and more ambitious action is needed to address it effectively. Damage caused by climate change has intensified in recent years, and its impacts are expected to become more severe. While mitigation efforts have gained momentum, they remain insufficient to meet global targets. Climate-related events, such as extreme heat and droughts, devastating floods and storms, and the spread of infectious diseases, are already affecting billions of people around the world. Since 1995, successive global summits and agreements, most prominently the 2015 Paris Agreement on Climate Change, have called for action to decarbonize the world's economies. Yet, in 2023, greenhouse gas (GHG) emissions were still rising, and global warming continues to accelerate. The first global stocktake of climate action under the Paris Agreement, conducted at the 28th Conference of the Parties (COP28) to the United Nations Framework Convention on Climate Change (UNFCCC) in November 2023, found a "glaring gap" in emissions reduction commitments and actions, with reviewers concluding that the world is "significantly off track" in meeting Paris Agreement targets (UN 2023).

The window to stay within the 1.5°C target of the Paris Agreement is rapidly closing. Parties to the Paris Agreement agreed to limit average global temperature rise to "well below 2°C above pre-industrial levels" and aim for 1.5°C. By 2023, the 1.5°C target had almost been exceeded, as average temperatures soared to 1.46°C above preindustrial averages (C3S 2024). Unless more ambitious and large-scale mitigation action is taken urgently, the world risks overshooting critical tipping points,[1] with disastrous consequences for the livability of large parts of the planet. As this report highlights, even if no tipping points are triggered, exceeding 2°C of warming would significantly increase risks, adverse impacts, and related damages and losses.

The transition to low-carbon economies is underway in Asia and the Pacific, but must be accelerated. Carbon intensity is declining steeply only in some parts of Asia, and it always declines as economies move to be more dominated by services. Developing Asia invests more in renewable energy than any other region in the world, and currently hosts more than 50% of the world's solar and wind generation capacity. Forty-four economies in the region have submitted nationally determined contributions (NDCs) to the UNFCCC, and many have adopted net zero targets.[2] Yet, emissions continue to rise from the region, and achieving net zero targets within the century will require large investments and transformative changes.

People in the region are highly concerned about climate change and its impacts. They also want to see more ambitious government action. In a 2024 online survey conducted by the Asian Development Bank (ADB), involving about 13,500 residents in 14 Asian economies, 91% of respondents identified climate change as a serious problem (Abiad et al. 2024).[3] As seen in Table 1.1, concern was highest in the subregions of South Asia (90%–92%) and Southeast Asia (84%–96%). Additionally, 84% of respondents from across the region reported that climate change was already affecting them or their families or that they expected to experience impacts within the next 10 years. However, awareness of government policies and pledges to

---

[1] Tipping points are defined by the Intergovernmental Panel on Climate Change (IPCC) as "critical thresholds in a system that, when exceeded, can lead to a significant change in the state of the system, often with an understanding that the change is irreversible." Examples of global-warming-related tipping points are the collapse of ice sheets in Greenland and Antarctica, which would result in rapid sea-level rise; the thawing of boreal permafrost, which would release large amounts of methane and contribute to rapid warming; and the disruption of ocean currents (ESA 2023).

[2] Net zero is defined as a state where emissions and removals of GHGs are balanced, meaning the economy no longer contributes to global warming (NZT 2024).

[3] ADB's climate change perception survey assessed about 13,500 respondents across 14 Asian economies and covered respondents' characteristics, perceptions of climate change, policy awareness, support for mitigation and adaptation measures, and willingness to adopt climate-friendly behaviors.

This chapter was written by a team of ADB staff consisting of Brendan A. Coleman, Jules Hugot, Arturo M. Martinez Jr., Homer Pagkalinawan, Madhavi Pundit (lead author), and Melanie Grace A. Quintos.

Table 1.1: Beliefs, Concerns, and Support for Policies on Climate Change
(% share of respondents)

	Share of Respondents	CCA	East Asia			South Asia				Southeast Asia						
		Kazakhstan	PRC	Hong Kong, China	Taipei,China	Bangladesh	India	Pakistan	Sri Lanka	Indonesia	Malaysia	Philippines	Singapore	Thailand	Viet Nam	Total
Perception	Climate change…is a serious problem	88	86	87	94	91	90	92	90	91	84	90	94	92	96	91
	… affects people now and in the next 10 years	87	82	82	91	90	94	89	93	89	87	90	92	88	98	90
	… affects my family now or within the next 10 years	75	78	74	85	86	87	86	83	86	79	86	86	86	92	84
	Government has taken action to reduce climate change	53	83	68	66	71	80	42	46	55	54	70	81	42	82	64
Concern about impacts	Heat waves	43	37	37	50	55	63	81	50	62	67	54	60	62	27	54
	Flooding	44	55	46	34	47	58	32	41	47	47	71	32	51	67	48
	Unpredictable weather	44	60	56	61	37	32	27	45	54	45	46	50	48	57	47
	Less productive agriculture/higher food prices/reduced food security	45	39	39	47	31	25	22	37	40	35	34	41	40	28	36
	Drought	46	46	36	41	37	21	17	34	47	22	21	29	57	43	35
Support for policies	Laws and regulations limiting emissions (e.g., emission standards for buildings and appliances)	57	61	62	63	44	67	36	59	44	47	43	60	36	60	53
	Investment in low-emission, resilient infrastructure (e.g., renewable energy, public transport)	74	71	71	70	56	75	52	69	56	63	59	68	50	76	65
	Carbon tax	54	65	50	55	53	74	42	64	37	47	45	53	46	68	54

CCA = Caucasus and Central Asia, PRC = People's Republic of China.
Notes: The figures in the table denote the share (%) of respondents by economy for each climate-related issue. Green denotes higher share; red lower share. Results are weighted by age, gender, location, and education.
Source: Abiad, A., J. R. Albert., A. Martinez, M. Pundit, and M. Thomas. 2024. Voices of Change: Public Sentiment on Climate Change Policies in Asia. Background paper for the *Asia–Pacific Climate Report 2024: Catalyzing Finance and Policy Solutions.* Asian Development Bank.

address climate change varied. It was relatively high in the People's Republic of China (PRC), Singapore, and Viet Nam, but notably lower in Pakistan, Sri Lanka, and Thailand. Most respondents supported mitigation and adaptation measures, especially investments in low-emission and resilient infrastructure, and flood protection.

1.2   Opportunities Arising from Low-Carbon Transition

Avoiding the worst consequences of climate change is still possible and necessary to prevent severe damage to communities and economies. Significant additional warming is already inevitable due to the long lifetime of carbon dioxide (CO_2) in the atmosphere, but action taken now can prevent the world from overshooting critical tipping points. This would require a whole of economy transition to low-carbon technologies and systems. Such transitions are particularly challenging in energy-intensive industry sectors such as steel and cement. Substantial changes would also be required in waste management, agriculture, and land-use practices, such as reduced deforestation and increased nature restoration and afforestation. While the transition to net zero emissions is anticipated to cost developing Asia a significant amount, the cost of inaction would be many times higher (ADB 2023a).

The co-benefits of climate action significantly outweigh the associated costs. These co-benefits include improved public health, reduced agricultural losses, and the creation of high-quality jobs, with the most significant advantages emerging in the medium to long term (although benefits are also notable in the short term). ADB (2023a) projects that the accelerated pursuit of net zero could generate 1.5 million jobs in the energy sector by 2050 and prevent up to 346,000 annual deaths from air pollution by 2030. Moreover, starting ambitious climate action now could reduce the costs of achieving net zero by 10%–20% compared to delaying until after 2030. Public support for climate action is growing as impacts become more apparent and as awareness increases about the potential to drive sustainable, inclusive economic growth. Scaling up climate finance to invest in advanced climate technologies and nature-based solutions can foster low-carbon, climate-resilient economic growth.

Developing Asia is well positioned to seize opportunities from climate action. Its comparative advantages include:

(i) **Green energy potential and pricing.** The region's economies have large renewable energy generation potential and can produce some of the world's cheapest low-carbon electricity. In 2022, solar photovoltaic electricity in the PRC and India was generated at $0.04 per kilowatt-hour (kWh), compared to $0.06/kWh in the United States and $0.08/kWh in Germany (OWD 2022). A similar trend is observed for onshore wind energy. The region's solar and wind generation capacity is also growing faster than anywhere else in the world (IRENA 2023). By developing a green hydrogen industry, the region could harness this affordable energy to support the industrial production of low-carbon products and materials, which are increasingly in demand.

(ii) **Capacity to transition to low-carbon production.** The region is home to some of the world's fastest-growing and most innovative economies (IMF 2024; WIPO 2023). Many economies in the region benefit from large workforces and high-quality technical education, providing significant opportunities to gain economic benefits from the global transition to low-carbon technologies. For example, economies in East Asia and Southeast Asia, with their strong manufacturing sectors, are well-placed to produce and export renewable energy components and other advanced technologies. In 2022, developing Asia manufactured 98.4% of the world's solar photovoltaic cells, with the PRC accounting for 84% and Malaysia, Thailand, and Viet Nam also contributing (IEA 2023). According to ADB (2023b), manufacturing related to renewable energy in Southeast Asia could generate between $90 billion and $100 billion in annual revenue by 2030 and create up to 6 million jobs by 2050.

(iii) **Scope for gains in carbon and energy efficiency.** The carbon intensity of the region's industry and energy sectors remains high compared to global averages. This indicates significant potential for efficiency gains using proven, cost-effective technologies through investments with short payback periods. By enhancing energy and carbon efficiency, the industry and energy sectors can reduce costs and boost competitiveness, while rapidly cutting GHG emissions and pollution.

Seizing these opportunities can deliver numerous benefits, including for vulnerable communities. Developing Asia has the potential to become a pivotal force in climate action by reducing emissions and pollution, while improving public and environmental health. These efforts will create millions of high-quality green jobs in sectors that are environmentally, socially, and economically sustainable. With well-designed policies, the transition to net zero also presents an opportunity to build more equitable and inclusive economies. This requires the development of growth strategies that incorporate inputs from, and address the needs of, all societal stakeholders—particularly the most vulnerable and traditionally underrepresented groups. By raising its ambitions now, developing Asia can lead the way and unlock significant benefits for the region.

1.3   Seizing the Moment for Climate Action

As the window to prevent severe climate impacts narrows, policymakers need reliable information. Information on climate trends and impacts is required for developing effective national climate strategies and for avoiding disorderly transitions and maladaptation. This report aims to provide actionable insights that support the development of well-informed climate policies and action plans aligned with global and national climate goals. In turn, these insights will facilitate the mobilization and deployment of climate finance with the greatest impact, ensuring optimal resource allocation for impactful climate solutions.

The current period is pivotal for advancing Paris Agreement goals. Starting in 2025, Parties to the UNFCCC will submit their updated NDCs, known as "NDCs 3.0". These updated commitments are expected to reflect more ambitious mitigation and adaptation targets following the first global stocktake at COP28 in 2023, which revealed insufficient progress. Many of ADB's developing member economies have announced net zero targets, but detailed road maps and sectoral plans have yet to be developed. Many economies also lack long-term strategies for decarbonization.

The report aims to engage all stakeholders by discussing the rationale and methods for ambitious climate action. It provides an overview of the latest data on emissions and climate impacts in developing Asia, highlights effective mitigation and adaptation policies and initiatives, explores equitable climate policies to garner broad public support, and discusses the role of climate finance in supporting these efforts.[4] While its primary goal is to facilitate evidence-informed policy dialogue and underscore the importance of setting ambitious climate targets, the report also informs the private sector, civil society, academia, and the public about climate risks, opportunities, adaptation initiatives, and mitigation pathways relevant to the region. Achieving decarbonization, adapting to climate change, and fostering a more inclusive and sustainable future require the active participation of all members of society.

1.4   Challenges for Developing Asia

The region both suffers from and contributes to climate change. Developing Asia is home to some of the world's most climate-vulnerable economies, where populations are already facing frequent extreme weather events, including intense heat waves, floods, and droughts. For example, the Lao People's Democratic Republic and Viet Nam recorded unprecedented high temperatures in 2023, while several low-lying small island economies in the Pacific are at risk of flooding due to rising sea levels. At the same time, the region includes three of the top 10 GHG-emitting economies globally—the PRC, India, and Indonesia.

Developing Asia has accounted for most of the increase in global GHG emissions since 2000. While advanced economies were major emitters of GHGs throughout the 20th century, emissions from developing Asia have grown more rapidly than those from any other region in the first 2 decades of the 21st century. Consequently, the region's share of global emissions rose from 29.4% in 2000 to 45.9% in 2021 (Figure 1.1). In contrast, Europe's emissions peaked before 2000 and those of the United States peaked in 2007 (Levin and Rich 2017). Emissions from developing Asia continue to rise, driven primarily by the PRC, which contributed about 30% of global emissions in 2021. The region is home to 60% of the world's population and though per capita emissions are still lower than the global average, effective climate action cannot be achieved without Asia's decarbonization efforts.

---

[4]   Additionally, ADB's *Key Indicators for Asia and the Pacific 2024: Data for Climate Action* provides a comprehensive assessment of data on climate change drivers, impacts, and vulnerability, along with strategies for mitigation and adaptation.

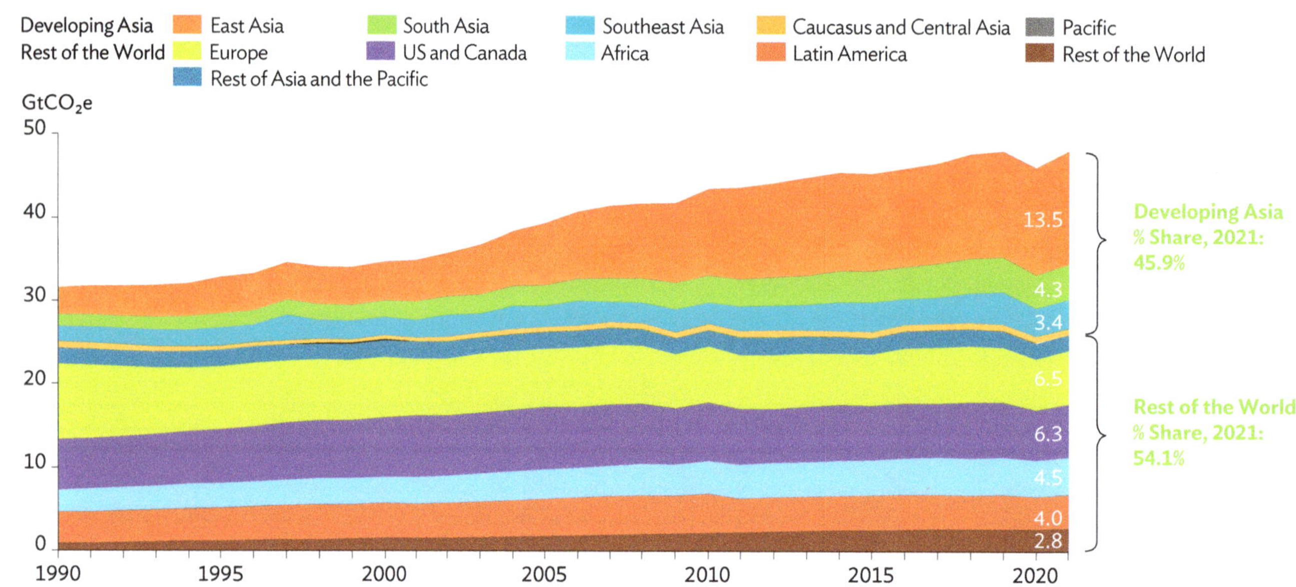

Figure 1.1: Greenhouse Gas Emissions in Selected Regions and Subregions of the World, 1990–2021

GtCO$_2$e = gigaton (1 billion metric tons) of carbon dioxide equivalent, US = United States.
Notes: "Developing Asia" excludes Hong Kong, China and Taipei,China due to lack of available data. Emissions from land-use change and forestry, which can be positive or negative, are included. Emissions per year calculations exclude economies with no available data.
Source: World Resources Institute. Climate Watch (accessed 23 August 2024).

Rising domestic consumption and increasing energy demand are key factors driving emissions in developing Asia. High production levels to meet the demands of advanced economies also contribute to this trend. While per capita emissions in most of the region are significantly lower than those in advanced economies, the PRC's per capita emissions have surpassed those of Europe and Japan, reaching about 9.05 tons of CO$_2$ equivalent in 2021 (Figure 1.2). Despite a rapid decline in carbon intensity since 2000, levels in developing Asia, particularly in East Asia, remain notably higher than those of advanced economies. Emissions intensity is expected to decrease further with the growth of the services sector, enhanced carbon and energy efficiency, and a shift from fossil fuels to renewable energy sources.

The energy sector is the largest source of emissions in developing Asia. Due to the region's heavy reliance on fossil fuels for energy production, its energy sector accounted for 77.6% of regional emissions in 2021 (Figure 1.3). For comparison, the global energy sector's contribution to worldwide emissions was 74.8%. Figure 1.4 highlights the region's contributions to global emissions across specific energy subsectors, showing that developing Asia generated a significant share in manufacturing and construction (63%) and in electricity and heat production (56%), with East Asia as the primary contributor. Although emissions in the transport subsector were lower (28%), they are projected to increase as economies develop and private vehicle ownership expands (ATO 2023).

Figure 1.2: Per Capita Emissions and Emissions Intensity in Developing Asia Versus Global Comparators

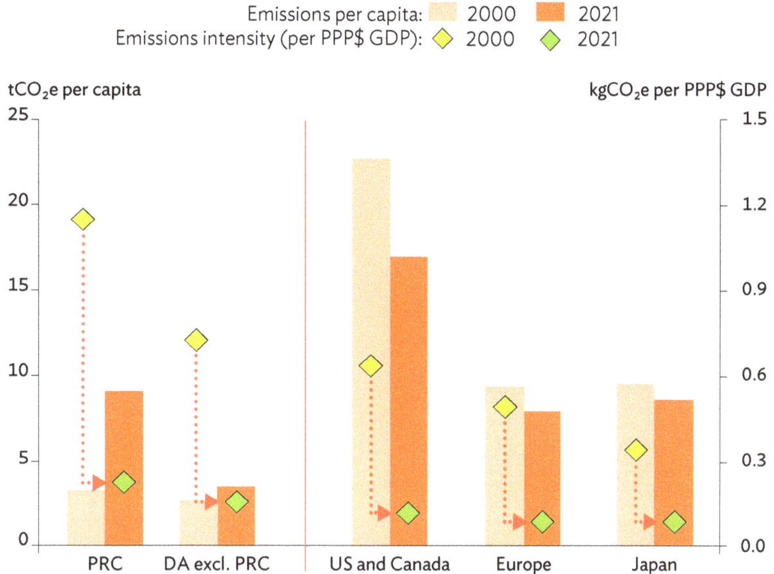

PRC = People's Republic of China, DA = developing Asia, GDP = gross domestic product, kgCO$_2$e = kilogram of carbon dioxide equivalent, PPP = purchasing power parity, tCO$_2$e = ton of carbon dioxide equivalent, US = United States.
Notes: Emissions from land-use change and forestry, which can be positive or negative, are included. Greenhouse gas emissions for developing Asia exclude Hong Kong, China and Taipei,China due to lack of available data. GDP PPP for developing Asia excludes the Cook Islands; Hong Kong, China; Niue; and Taipei,China. Global emissions per year calculations exclude economies with no available data. Global GDP PPP per year calculations exclude economies with no available data.
Sources: World Resources Institute. Climate Watch (accessed 23 August 2024); International Monetary Fund. World Economic Outlook Database (accessed 23 August 2024); and CEIC Data Company.

Figure 1.3: Emissions by Sector in Developing Asia, 2021

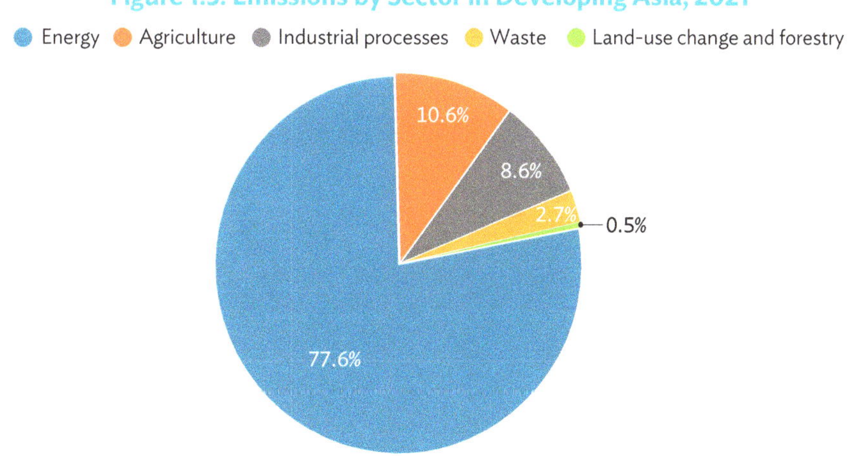

Source: World Resources Institute. Climate Watch (accessed 23 August 2024).

Figure 1.4: Share of Regional Emissions in Total Global Emissions for 2021, by Energy Subsectors

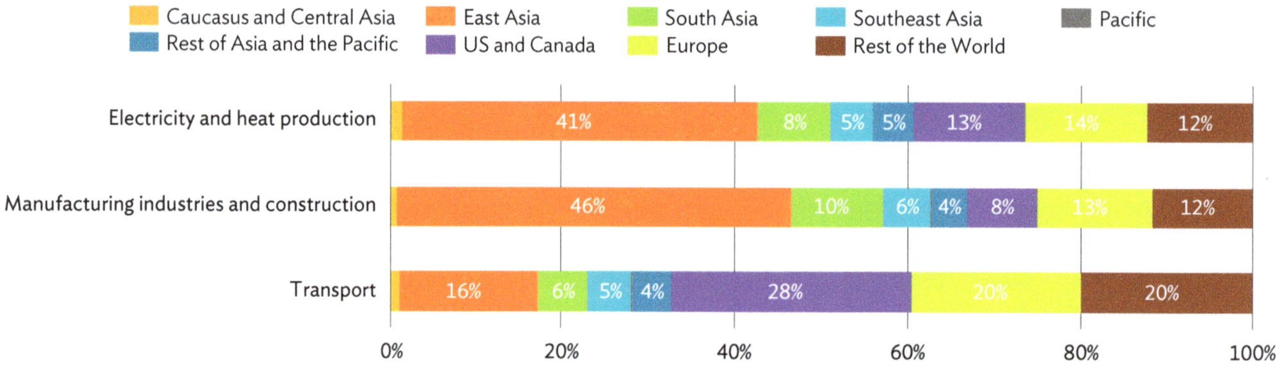

US = United States.
Note: Greenhouse gas emissions excludes economies with no available data, e.g., Hong Kong, China and Taipei,China.
Source: World Resources Institute. Climate Watch (accessed 23 August 2024).

The high carbon intensity of the region's electricity sector poses both challenges and opportunities. While the sector holds significant potential for rapid decarbonization, there is a risk of locking in high emissions for decades if fossil-fuel-based power plants are not decommissioned promptly. To mitigate this risk, governments in developing Asia should consider accelerating decommissioning timelines. Additionally, since renewable electricity generation is often more cost-effective than fossil-fuel-based options, it is crucial for economies to ensure that any new installed generation capacity is low carbon.

Investment in renewable energy is strong and this momentum should be sustained. In 2023, foreign direct investment in sustainable development projects—mainly in renewable energy, transport, and telecommunications—increased by 54% in developing Asia (UNCTAD 2024). The region has emerged as a global leader in wind and solar power, home to 52.5% of the world's installed capacity in 2022, largely due to significant investments in the PRC, which accounted for over 73% of this capacity (ZCA 2023). In the Caucasus and Central Asia, Southeast Asia, and the Pacific, hydropower remains the dominant renewable energy source, relative to total capacity (Figure 1.5). Several smaller economies including Bhutan, Nepal, the Kyrgyz Republic, and Tajikistan, generate the majority of their electricity (86%–100%) through hydropower. To achieve net zero emissions, substantial investments are needed not only in generation capacity but also in energy storage, grid enhancement,

Figure 1.5: Share of Renewables in Total Installed Capacity in 2023, by Subregion

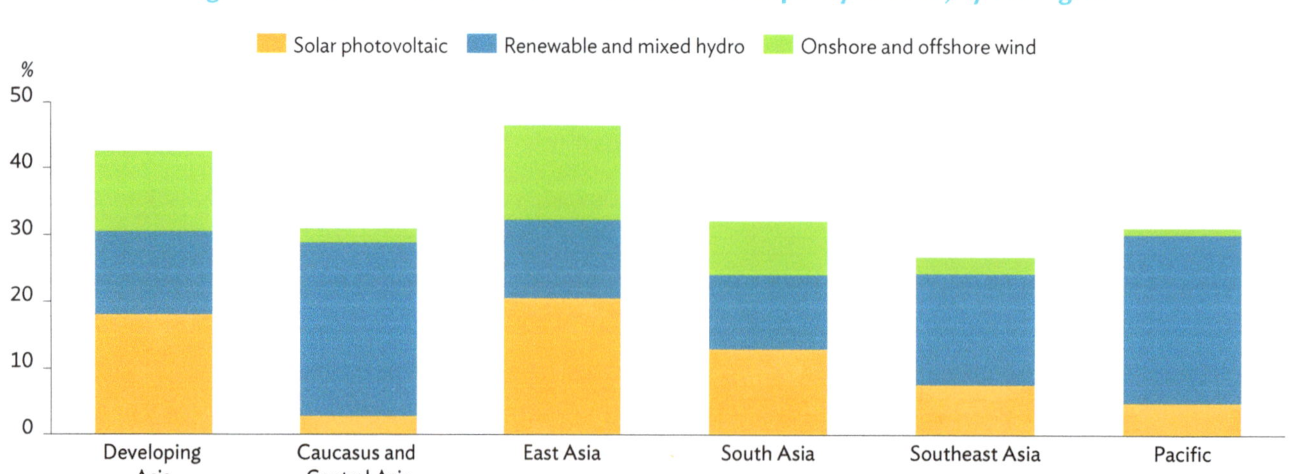

Source: International Renewable Energy Agency. IRENA Database (accessed 8 October 2024).

and smart grid technologies to manage the intermittent nature of sources such as wind and solar energy. Meanwhile, the extensive hydropower resources in developing Asia can support grid stability as wind and solar power capacity increase, and enhanced grid interconnection can help to further mitigate intermittency challenges.

As the cost of renewable energy technology continues to decline, capacity is expected to expand further across developing Asia. Manufacturing costs of solar photovoltaic cells and batteries are lower in developing Asia than in other parts of the world, and still decreasing. This competitiveness has driven the rapid expansion of renewable generation capacity in the region. In 2022, the PRC and India achieved some of the lowest solar photovoltaic installation costs globally, while generation across developing Asia more than doubled from 2018 to 2022 (Figure 1.6). Although less pronounced, a similar pattern has also been observed in wind turbine installations. The connected trends of growing capacity and declining prices are expected to continue, as lower costs for energy storage technologies will further boost the competitiveness of renewable energy and enable more diverse applications, especially in industrial electrification, transportation, and heating.

Figure 1.6: Renewable Energy Installation Costs and Generation

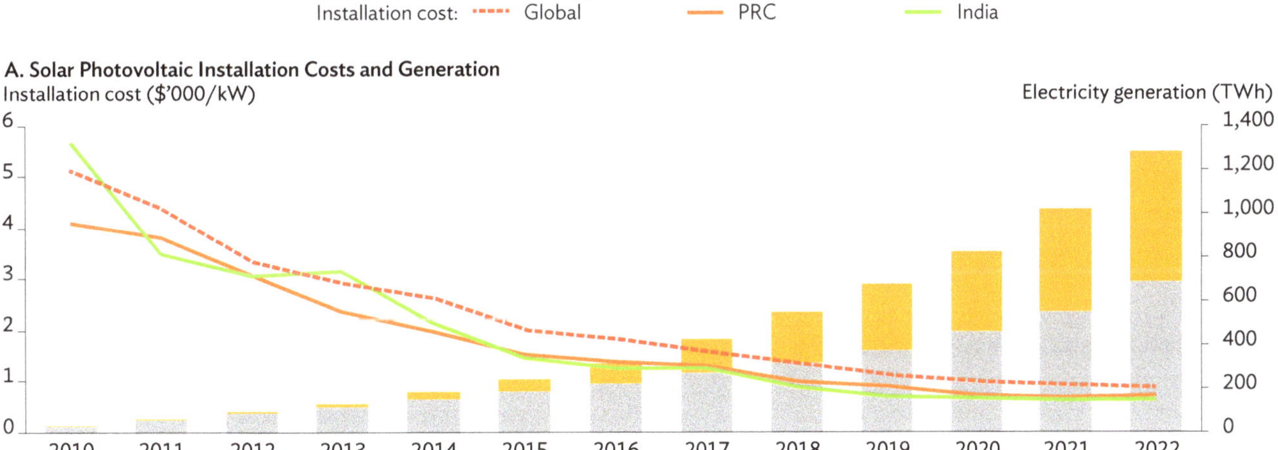

PRC = People's Republic of China, kW = kilowatt, TWh = terawatt-hour.
Source: International Renewable Energy Agency. IRENA Database (accessed 23 August 2024).

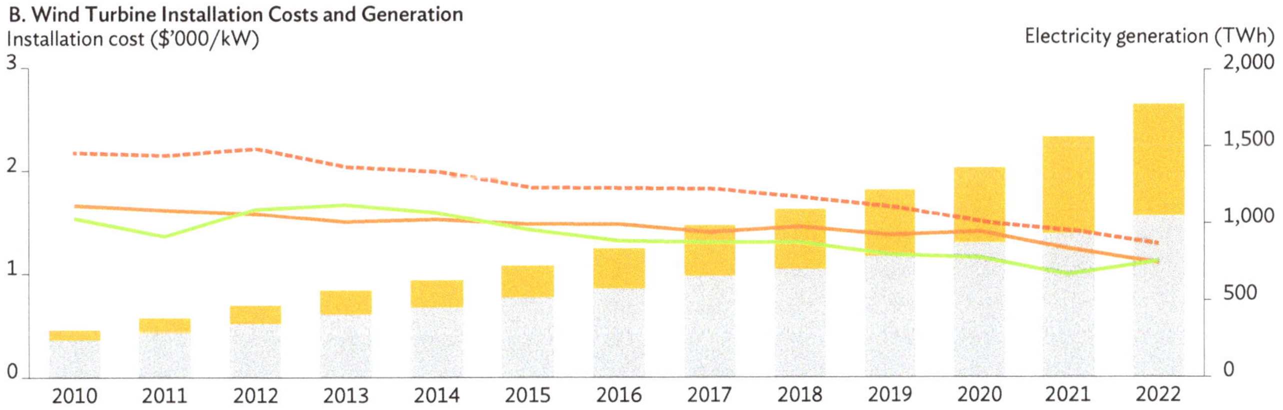

PRC = People's Republic of China, kW = kilowatt, TWh = terawatt-hour.
Source: International Renewable Energy Agency. IRENA Database (accessed 10 September 2024).

Agriculture and industrial processes also contribute significantly to GHG emissions in the region.
In 2021, agricultural emissions accounted for 10.6% of the region's total emissions, at 2.3 gigatons of carbon dioxide equivalent ($GtCO_2e$), while industrial emissions made up 8.6% (at 1.9 $GtCO_2e$). Key drivers of agricultural emissions include the livestock sector and methane emissions from rice cultivation. Improved rice cultivation practices, enhanced livestock management, and sustainable farming techniques should be explored for their potential to reduce emissions rapidly and at low cost (IPCC 2019; Rahut et al. 2023).

Emissions from land-use change and forestry are substantial in some economies. In Southeast Asia, particularly in Indonesia, deforestation for palm oil, timber, and agriculture is a major source of emissions (Chen et al. 2024). Meanwhile, Bhutan's extensive forests have helped it achieve net zero, which highlights the need for efforts to reduce deforestation and enhance forests as carbon sinks across developing Asia. Several economies, including the PRC, India, Indonesia, Malaysia, and the Philippines, have set ambitious reforestation targets to mitigate climate change and protect biodiversity (Bosshard 2022; WEF 2022).

The region's transport emissions, although low as a share of total regional emissions, are rising and require innovative solutions. As of 2021, developing Asia accounted for 28% of global transport emissions, up from 14% at the turn of the century. Since access to transport remains relatively limited in the region, particularly in rural areas, the sector's growth is necessary to spur economic development. However, combined with urbanization, this growth could lead to transport emissions more than doubling from 2.9 $GtCO_2e$ in 2018 to 7 $GtCO_2e$ in 2050 (ATO 2023). To mitigate increases in transport emissions, economies should promote alternatives such as two- and three-wheeled vehicles over passenger cars while also developing integrated mass transit systems.

1.5   The Need for Adaptation

Climate change will drastically alter the living conditions for people in developing Asia. The most pronounced impacts of climate change in the region will be felt in the frequency, duration, and severity of heat waves, followed by shifts in precipitation patterns that will elevate flood risks and increase drought occurrences. More severe storms and rising sea level are expected to increasingly threaten low-lying coastal areas, including coastal megacities where populations and economic assets are heavily concentrated. Chapter 2 quantifies the economic implications of a high end emissions scenario, estimating GDP reductions of 17% by 2070, which may increase to 41% by 2100.

People are increasingly recognizing the severity of threats to the region posed by climate change.
Respondents to ADB's climate change perception survey expressed the highest concern about severe heat waves (54%), floods (48%), and unpredictable weather (47%), mirroring scientists' expectations of climate impacts (Table 1.1). The survey revealed that climate change concerns varied depending on the lifestyles and demographics of respondents. Users of public transportation and those exposed to diverse news sources showed higher concern; and higher education levels generally correlated with increased concern. Age effects varied by economy: individuals aged 35 to 54 were more concerned in some economies, such as the Philippines and Singapore; while those aged 55 and older were more concerned in other economies, including India, Kazakhstan, and Sri Lanka. These findings underscore the complex factors that shape perceptions and highlight the need for targeted climate communications and policy strategies.

Increased investment in climate change adaptation initiatives is essential. This investment can help reduce the exposure and vulnerability to climate risk of developing Asia's communities and economies. By enhancing resilience, these initiatives can protect critical infrastructure, safeguard livelihoods, and ensure food and water security in the face of climate challenges. However, this report highlights that the gap between required and available adaptation finance is substantial, potentially reaching hundreds of billions of dollars annually. To address this, governments will need to explore new sources of funding for adaptation initiatives.

Residents of developing Asia are also calling for increased investment in resilience and adaptation. ADB's climate change perception survey highlights a strong consensus among respondents on the need for robust flood protection infrastructure, early warning systems, and climate-resilient infrastructure[5] (Figure 1.7). Measures such as resilient water supply, coastal protection, agricultural adaptation, and conservation of natural areas also received support in the survey, although priorities may vary by local context. For financing adaptation initiatives, over half of the respondents favored efforts to reduce corruption and tax evasion, as well as implementation of progressive taxation. Other preferred financing options included international support through climate funds and foundations, and, to a lesser extent, carbon taxation. These preferences underscore the importance of establishing equitable and efficient financing mechanisms to secure broad support for investment in adaptation initiatives.

Figure 1.7: Public Preferences for Government Adaptation Measures and Funding Options

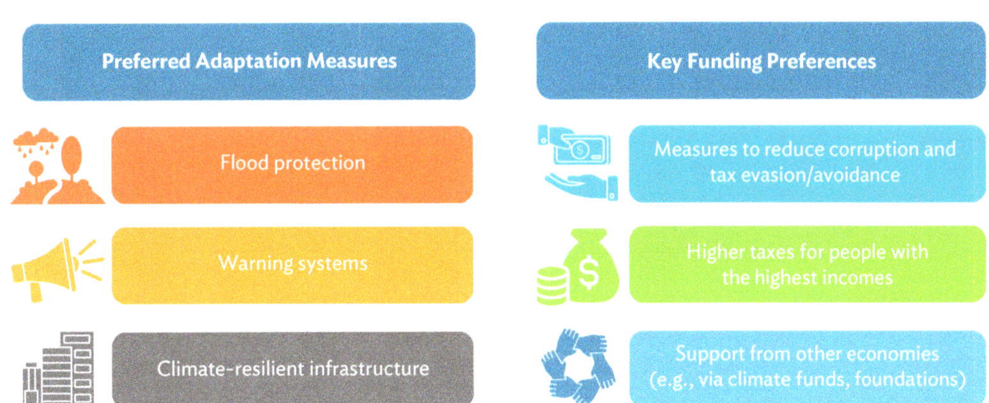

Source: Abiad, A., J. R. Albert., A. Martinez, M. Pundit, and M. Thomas. 2024. Voices of Change: Public Sentiment on Climate Change Policies in Asia. Background paper for the *Asia Pacific Climate Report 2024: Catalyzing Finance and Policy Solutions*. Asian Development Bank.

1.6 Scope for Higher Ambitions and More Concrete Plans

Accelerating mitigation is key to avoiding the worst impacts of climate change, and feasible pathways for rapid decarbonization exist. ADB (2023a) highlights that key opportunities for mitigating emissions in developing Asia currently lie in increasing energy efficiency and changing land use practices. Under the accelerated global net zero scenario, improvements in energy efficiency are projected to be the most significant source of emission reductions prior to 2040, followed by non-CO_2 abatement, particularly in agriculture and land use. In the long term, most mitigation is achieved by changing the energy mix to be dominated by renewables and use of carbon capture and storage, with the power sector transitioning faster than the rest of energy. While transitioning from fossil fuel-based generation to renewable energy takes time, adopting energy-efficient behaviors and equipment can happen more swiftly. In the case of Southeast Asia, land use emissions can be rapidly reduced by halting deforestation and making simple changes to reduce methane emissions from rice production. Even under moderate climate action scenarios, the share of coal in the region's energy mix is expected to decrease from nearly half in 2023 to less than a quarter by 2050. With more ambitious policies, this share of coal could drop to 13% during the same period. Table 1.2 outlines further milestones under the accelerated global net zero scenario that economies in developing Asia should aim for between now and 2070 to align with global climate goals.

---

[5]   Climate-resilient infrastructure is infrastructure that is planned, designed, built, and operated with changing climate impacts in mind (Cho 2024).

Table 1.2: Select Key Milestones to 2070 Under the Accelerated Global Net Zero Scenario

Year	Energy	Carbon Price	Land Use	Co-Benefits
2030		The carbon price reaches $70 per ton of CO_2e.	Improved air quality prevents crop yield losses of 8 million tons.	346,000 premature deaths from air pollution are prevented annually.
2035	Coal is almost completely phased out of the electricity mix.			
2040	Wind and solar provide 75% of the region's electricity.			
2050	Renewable energy provides 40% of the region's energy needs. Installation costs of solar photovoltaic have declined by 76% and of onshore wind by 66%. Fossil fuel imports are reduced by 71% in South Asia, 45% in Southeast Asia, and 36% in India. In the PRC, more than 25% of final energy in road transport is provided by electricity.	The carbon price reaches $153 per ton of CO_2e.	Forest cover has increased by 95 million ha to 30% of the total area; 39 million ha of land is used to grow energy crops.	1.4 million coal jobs are lost, but 2.9 million new renewable energy jobs are created.
2070	Renewable energy sources provide 50% of the region's energy needs. Per capita electricity consumption has increased 2.7 times since 2020; the share of electricity in final energy consumption is over 50%. 87% of final energy in road transport is provided by electricity.		97 million ha of land (5% of land area) is used to grow energy crops.	

PRC = People's Republic of China, CO_2e = carbon dioxide equivalent, ha = hectare.
Source: Asian Development Bank. 2023. *Asia in the Global Transition to Net Zero: Asian Development Outlook 2023 Thematic Report.*

Governments in developing Asia are already working to combat climate change, but there is scope to further raise ambitions. Within developing Asia, 44 economies (which account for 98% of the region's total emissions) as Parties to the UNFCCC, have ratified the Paris Agreement, and have submitted their NDCs. These NDCs outline commitments to reduce national GHG emissions and adapt to climate impacts, with some detailing strategies for finance, technology, and capacity building. Since 2020, all Parties have revisited, and some have increased, their NDC ambitions. For example, India has achieved two of its NDC targets ahead of time and has increased its goal to reduce the emissions intensity of its GDP from 35% to 45% by 2030, compared to a 2005 baseline. Viet Nam raised its emissions reduction target for 2030 from 9% to 15.8% (unconditional) and from 27% to 43.5% (conditional).[6] Viet Nam also committed to reducing methane emissions by 30% by 2030, aligning with the Global Methane Pledge,[7] and developed a Methane Emission Reduction Action Plan through to 2030 (UNFCCC 2022). Nepal's second NDC submission provides more detail than the first, with quantified targets for agriculture, energy, forestry, and waste, and a commitment to gender equality and social inclusion by empowering women, Indigenous Peoples, and youth in climate policy (MHP 2020). Despite these updates, NDC commitments across developing Asia remain insufficient to align with the Paris Agreement goals (Figure 1.8).

---

[6]   Unconditional targets are those that a country commits to achieving independent of additional external financial support. Conditional targets can only be achieved with additional external financial support.

[7]   Methane is a highly potent GHG with a shorter atmospheric lifetime than CO_2. Its warming potential is 28 times greater than CO_2 over a 100-year period and 84 times greater over a 20-year period (European Commission n.d.).

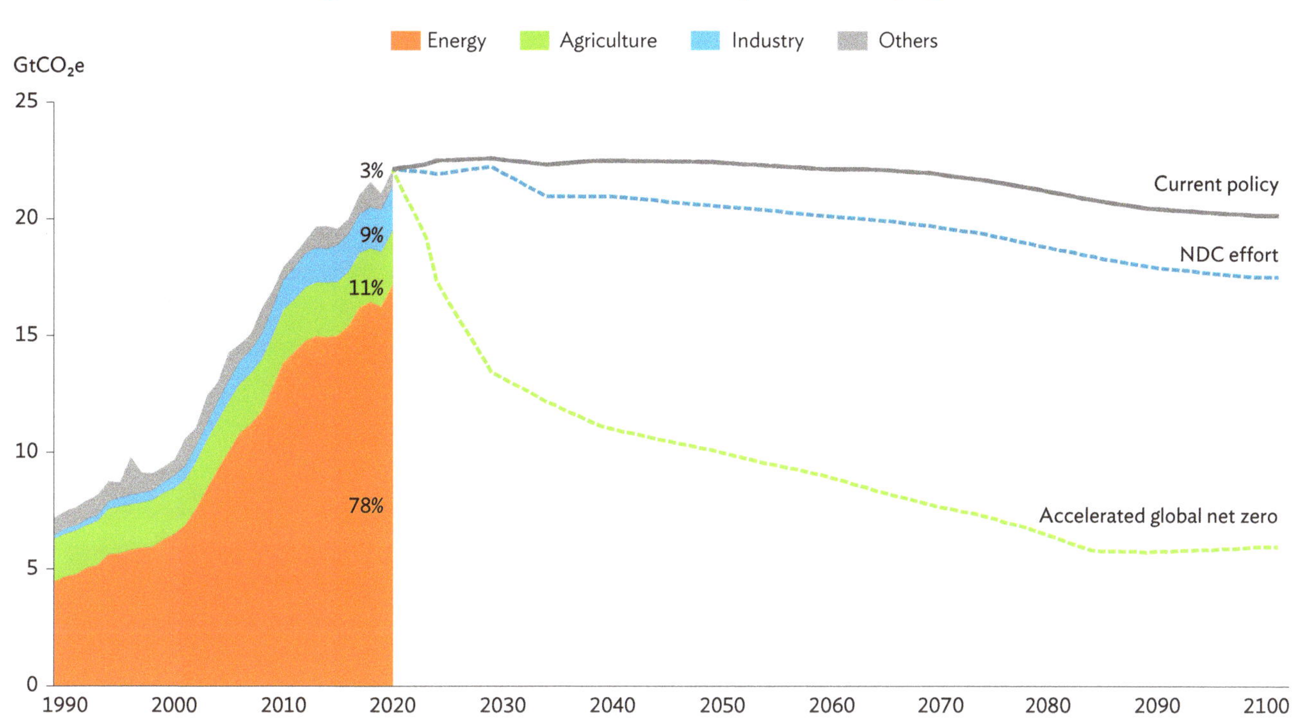

Figure 1.8: Possible Emissions Pathways for Developing Asia

GHG = greenhouse gas, GtCO$_2$e = gigaton (1 billion metric tons) of carbon dioxide equivalent, NDC = nationally determined contribution.
Notes: GHG emissions for developing Asia exclude Hong Kong, China and Taipei,China due to lack of available data. Scenarios are generated from the World Induced Technical Change Hybrid model; and show GHG pathways for developing Asia under current policies, NDC effort, uncoordinated net zero, global net zero, and accelerated global net zero.
Sources: World Resources Institute. Climate Watch (accessed 23 August 2024); and Asian Development Bank. 2023. *Asia in the Global Transition to Net Zero: Asian Development Outlook 2023 Thematic Report*.

The comprehensiveness of NDCs in the region varies significantly. Among the 44 economies of developing Asia with submitted NDCs, all include targets for the energy sector, but only 40 include targets for the agriculture, forestry, and other land use sector (Figure 1.9). Moreover, just 30 economies have targets for the industry sector despite the crucial role this sector plays in decarbonization. An analysis of the climate mitigation measures outlined in the 44 updated NDCs, based on the UNFCCC typology (2023b), highlights both priority sectors and gaps in climate planning within the region.

Renewable energy generation is the mitigation action most frequently mentioned in NDCs, but other critical actions are lacking. Measures to improve the energy efficiency of buildings are also commonly mentioned, but key actions within the agriculture, forestry, and other land use sector—such as reducing deforestation, restoring land, and promoting agroforestry—are included in fewer than half of the NDCs. Measures for the transport sector primarily focus on electrification and developing efficient vehicles, rather than on mode shifting and demand reduction. Crosscutting measures and the promotion of a circular economy appear in relatively few NDCs, indicating a lack of holistic and integrated approaches to climate action and sustainable development. Similar patterns are evident in the long-term strategies outlining 2050 goals.

Figure 1.9: Frequently Mentioned Mitigation Options in the Nationally Determined Contributions of Developing Asia

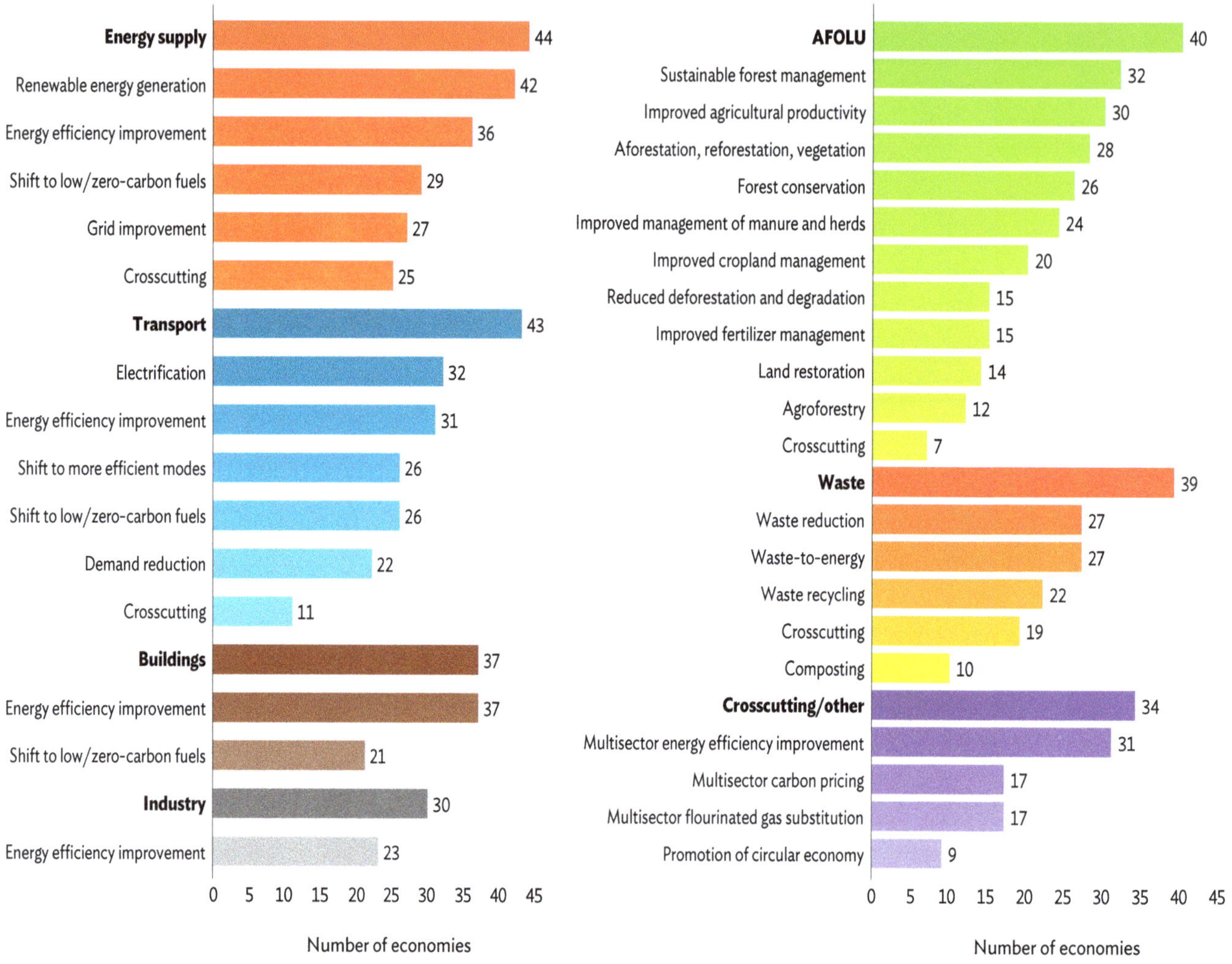

AFOLU = agriculture, forestry, and other land use.
Note: Compiled by the Asian Development Bank based on a review of the nationally determined contributions of 44 economies in developing Asia following the typology used by the United Nations Framework Convention on Climate Change in 2023.
Source: United Nations Climate Change. Nationally Determined Contributions Registry (accessed 26 August 2024).

Setting ambitious goals is not enough; they must be followed by effective implementation. According to the UNFCCC, if the latest NDCs are fully implemented worldwide, emissions in 2030 will be approximately 8.8% above 2010 levels. This represents a slight improvement from the 2023 assessment, which projected emissions would be 10.6% above 2010 levels. Analysis also suggests that global emissions are likely to peak sometime in the 2020s, largely due to anticipated peaks in emissions from the PRC, with 2030 emissions expected to be 2% below 2019 levels (UNFCCC 2023a). However, there remains a significant gap between NDC commitments and their actual implementation, due primarily to inadequate policies, insufficient funding, and a lack of comprehensive monitoring and enforcement mechanisms (UNEP 2023b). To achieve meaningful emissions reductions, it is crucial to both ramp up the ambitions within NDCs and ensure their effective implementation.

There is scope for developing Asia to strengthen its net zero goals, recognizing that progress is needed on a global scale. Of the 44 economies in developing Asia with submitted NDCs, 36 have set targets for achieving net zero GHG emissions. According to the Net Zero Tracker, 28 of these economies aim for a 2050 target, while the PRC and India have set their targets for 2060 and 2070, respectively (Table 1.3). Nearly half (15) of the region's economies with net zero goals have established net zero policies, but only four have enshrined their targets in law. Meanwhile, 13 economies have proposed targets but have not yet adopted them, and three have made pledges that remain unformalized. Bhutan claims to have achieved net zero and must maintain this status amid expanding economic activity. Detailed net zero plans are lacking in 11 economies and incomplete in 22, with only three having fully developed plans in place. In contrast, 32 of 38 Organisation for Economic Co-operation and Development (OECD) economies have set net zero or climate neutrality targets, with 23 enshrined in law, and nearly all aiming for 2050, including six northern and central European economies committed to achieving net zero prior to 2050. To enhance net zero ambitions, developing Asia needs more policy guidance and climate action financing. Multilateral institutions such as ADB can play a key role in providing this support and helping to solidify and implement climate goals.

Table 1.3: Status of Net Zero Targets in Developing Asia

Economy	Target	End Date	Target Status	Plan
Caucasus and Central Asia				
Armenia	Net zero	2050	Pledge	Completed
Azerbaijan			No net zero	
Georgia	Net zero	2050	In policy document	Incomplete
Kazakhstan	Net zero	2060	In law	Incomplete
Kyrgyz Republic	Net zero	2050	Proposed	No
Tajikistan[a]			No net zero	
Turkmenistan			No net zero	
Uzbekistan[b]			No net zero	
East Asia				
People's Republic of China	Net zero	2060	In policy document	Incomplete
Republic of Korea	Net zero	2050	In law	Incomplete
Mongolia[c]			No net zero	
South Asia				
Afghanistan	Net zero	2050	Proposed	No
Bangladesh	Net zero	2050	Proposed	No
Bhutan	Net zero	2030	Achieved (self-declared)	Incomplete
India	Net zero	2070	In policy document	No
Maldives	Net zero	2030	In law	No
Nepal	Net zero	2045	In policy document	No
Pakistan	Net zero	2050	Proposed	No
Sri Lanka	Net zero	2050	Pledge	Incomplete

continued on next page

Table 1.3 *continued*

Economy	Target	End Date	Target Status	Plan
Southeast Asia				
Brunei Darussalam[d]	No net zero			
Cambodia	Net zero	2050	In policy document	Completed
Indonesia	Net zero	2060	Proposed	Incomplete
Lao People's Democratic Republic	Net zero	2050	In policy document	Incomplete
Malaysia	Net zero	2050	In policy document	Incomplete
Myanmar	Net zero	2050	Proposed	Incomplete
Philippines	No net zero			
Singapore	Net zero	2050	In policy document	Incomplete
Thailand	Net zero	2065	In policy document	No
Timor-Leste	Net zero	2050	Proposed	Incomplete
Viet Nam	Net zero	2050	In policy document	Incomplete
Pacific[e]				
Fiji	Net zero	2050	In law	Completed
Kiribati	Net zero	2050	Proposed	No
Marshall Islands	Net zero	2050	In policy document	Incomplete
Federated States of Micronesia	Net zero	2050	Pledge	No
Nauru	Net zero	2050	Proposed	Incomplete
Niue	Net zero	2050	Proposed	No
Palau	Net zero	2050	Proposed	Incomplete
Papua New Guinea	Net zero	2050	In policy document	Incomplete
Samoa	Net zero	2050	Proposed	Incomplete
Solomon Islands	Net zero	2050	In policy document	Incomplete
Tonga	Net zero	2050	Proposed	Incomplete
Tuvalu	Net zero	2050	In policy document	Incomplete
Vanuatu	Net zero	2050	In policy document	Incomplete

[a] The Government of Tajikistan announced in December 2023 its goal to become a "green country" by 2037.
[b] Uzbekistan's Ministry of Energy signed a Memorandum of Understanding in 2021 with the European Bank for Reconstruction and Development (EBRD) to cooperate towards achieving net zero by 2050.
[c] Mongolia's President pledged a Net Zero target to be achieved by 2050 at COP 27 World Leaders Summit in 2022.
[d] Brunei Darussalam's Minister of Development and co-chair of the National Council on Climate pledged to "move towards" net zero by 2050 in a high-level statement at COP26 in 2021.
[e] Cook Islands' Prime Minister committed to net zero emissions by 2040 in a national statement at COP 27 in 2022.
Notes: A plan is considered complete if it includes (i) measures for all emission scopes covered by the target; (ii) expected emission reductions from these measures within a specified time period; (iii) the extent of measure implementation; and (iv) a schedule for regular reviews, as outlined in the Net Zero Tracker. A plan is deemed incomplete if it includes some but not all of this information, while 'No Plan' indicates that no plan exists.
Sources: Net Zero Tracker. Data Explorer (accessed 23 August 2024); Government announcements.

Planning for climate change adaptation also requires more attention. In 2023, Bangladesh, Bhutan, the Marshall Islands, Pakistan, and Papua New Guinea submitted National Adaptation Plans (NAPs) to the UNFCCC, bringing the total number of developing Asia's economies with NAPs to 15. These plans are crucial for leveraging international climate finance: by late 2023, the Green Climate Fund had provided $5.3 billion to 83 adaptation projects linked to 44 submitted NAPs (UNCC 2023). However, fewer than half the region's economies have included calculations of their adaptation finance needs in either their NDCs or NAPs (UNEP 2023).

Addressing institutional, governance, and financing challenges is critical for effective implementation of climate strategies and plans. Additional barriers include limited access to technology, insufficient technical capacity and expertise, and vulnerability to climate events that divert resources away from decarbonization. Inadequate international climate finance and limited access to climate funds also impede progress. Across Asia and the Pacific, an estimated $2 trillion annual investment is needed to meet NDC targets during 2022–2030 (ADB 2023c). This investment demand far exceeds the annual global climate finance of $1.27 trillion during 2021–2022 (CPI 2023). To bridge the gap between current commitments and actual needs, it is essential to conduct a thorough assessment of these barriers and explore solutions tailored to each economy's context, while enhancing capabilities for climate-responsive fiscal planning.

1.7 Supporting a Just, Inclusive, and Comprehensive Transition

A successful transition to net zero requires a just transition approach.[8] This means the economic and social impacts of transitioning to a low-carbon economy are distributed equitably, thereby preventing increased inequality and fostering broad public support for climate action. However, not all governments have established comprehensive strategies to ensure that the public, particularly vulnerable communities and underrepresented groups, are consulted and engaged in climate action. While gender considerations are increasingly included in NDCs as an aspect of climate action, they are often addressed in a limited way, for example, focusing on women's vulnerabilities or household roles rather than their underrepresentation in political decision-making. There is also a need for greater focus on private sector engagement in shaping and implementing climate strategies, while the broader concept of just transition should be emphasized to address the needs of all stakeholders, including communities affected by economic changes (Figure 1.10).

Figure 1.10: Number of Nationally Determined Contributions That Mention Key Societal Issues

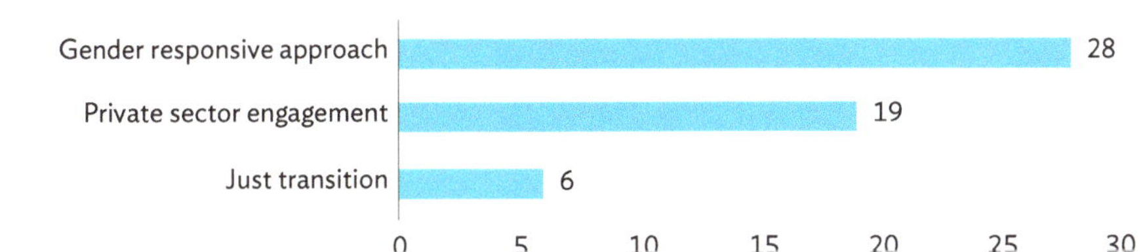

Note: Compiled by the Asian Development Bank based on a review of the nationally determined contributions of 44 economies in developing Asia.
Source: United Nations Climate Change. Nationally Determined Contributions Registry (accessed 26 August 2024).

---

[8]   The International Labour Organization defines "just transition" as "a process by which economies that progress toward a green economy also strengthen each of the four pillars of decent work for all (i.e., social dialogue, social protection, rights at work, and employment)."

The low-carbon transition entails a comprehensive, economy-wide transformation, that hinges on broad public support. The climate change perception survey identified key factors that affect both public support for climate policies and willingness to adopt climate-friendly behaviors. These factors include individuals' perceptions of how climate change directly impacts them, their awareness of climate policies, and demographic variables such as age, gender, and income. The survey also highlighted significant variations across economies, underscoring the importance of local contexts in shaping climate-friendly behaviors. While there is support for climate measures across developing Asia, skepticism about their effectiveness and costs persists. Increasing transparency in policy design and impact monitoring, ensuring an equitable distribution of costs and benefits, and closing awareness gaps through tailored communication that considers local contexts, and diverse demographics can address this skepticism and boost public support.

Developing Asia must create tailored mitigation pathways to effectively decarbonize. This involves implementing policies that leverage each economy's unique strengths while aligning with their socioeconomic objectives and development goals. A comprehensive transition is needed to address climate change, encompassing energy systems, infrastructure, policy, and financing, among other areas. Collaborative efforts among governments, the private sector, and communities will be essential to ensure that these pathways are both practical and sustainable, ultimately supporting climate resilience and economic growth.

References

Abiad, A., J. R. Albert., A. Martinez, M. Pundit, and M. Thomas. 2024. *Voices of Change: Public Sentiment on Climate Change Policies in Asia*. Background paper for the *Asia–Pacific Climate Report 2024: Catalyzing Finance and Policy Solutions*. Asian Development Bank.

ADB (Asian Development Bank). 2023a. *Asia in the Global Transition to Net Zero: Asian Development Outlook 2023 Thematic Report.*

_____. 2023b. *Renewable Energy Manufacturing: Opportunities for Southeast Asia.*

_____. 2023c. *Climate Finance Landscape of Asia and the Pacific.*

_____. 2024. *Key Indicators for Asia and the Pacific 2024: Data for Climate Action.*

ATO (Asian Transport Outlook). 2023. *2023 Climate Tracker for Transport in Asia & Pacific.*

Bosshard, E. 2022. Ambitious Tree Planting Goals in Asia Lack Diverse Tree Seeds from Native Species. *The Conversation.* 14 January.

C3S (Copernicus Climate Change Service). 2024. *Copernicus: February 2024 Was Globally the Warmest on Record – Global Sea Surface Temperatures at Record High.* 5 March.

Chen, S., C. Woodcock, L. Dong, K. Tarrio, D. Mohammadi, and P. Olofsson. 2024. Review of Drivers of Forest Degradation and Deforestation in Southeast Asia. Remote Sensing Applications: Society and Environment. 33. 101129.

Cho, R. 2024. The Case for Climate-Resilient Infrastructure. Columbia Climate School. 22 July.

CPI (Climate Policy Initiative). 2023. *Global Landscape of Climate Finance 2023.*

European Commission. n.d. Methane Emissions.

ESA (European Space Agency). 2023. Understanding Climate Tipping Points. 6 December.

ICCT (International Council on Clean Transportation). 2023. *Vision 2050 Strategies to Align Global Road Transport with Well Below 2°C.*

IEA (International Energy Agency). 2023. *Trends in Photovoltaic Applications.*

IMF (International Monetary Fund). 2024. IMF DataMapper: Real GDP Growth (accessed 23 August 2024).

IPCC (Intergovernmental Panel on Climate Change). 2019. *Climate Change and Land.* P. R. Shukla, J. Skea, E. Calvo Buendia, V. Masson-Delmotte, H.-O. Pörtner, D. C. Roberts, P. Zhai, et al., eds. Cambridge University Press. p. 896.

IRENA (International Renewable Energy Agency). 2023. Renewable Power Generation Costs in 2022.

_____. n.d. IRENASTAT Online Data Query Tool (accessed 23 August 2024).

Levin, K. and D. Rich. 2017. Turning Point: Which Countries' GHG Emissions Have Peaked? Which Will in the Future? World Resources Institute. 2 November.

MHP (Government of Nepal, Ministry of Health and Population). 2020. *Second Nationally Determined Contribution (NDC).*

NZT (Net Zero Tracker). 2024. Data Explorer (accessed 23 August 2024).

OWD (Our World in Data). 2022. Levelled Cost of Energy by Technology, 2022.

Rahut, D., D. Narmandakh, X. He, Y. Yao, and S. Tian. 2023. *Opportunities and Challenges During Low-Carbon Transition: A Perspective from Asia's Agriculture Sector.* ADB.

UN (United Nations). 2023. Global Stocktake Reports Highlight Urgent Need for Accelerated Action to Reach Climate Goals.

UNCC (United Nations Climate Change). 2023. Record Number of National Adaptation Plans Submitted in 2023, But More Are Needed. 13 December.

UNCTAD (United Nations Trade and Development). 2024. Developing Asia: Large Jump in Greenfield Investment in 2023.

UNEP (United Nations Environment Programme). 2023a. *Adaptation Finance Gap Update 2023.*

_____. 2023b. *Emissions Gap Report.*

UNFCCC (United Nations Framework Convention on Climate Change). 2022. *Viet Nam NDC 2022 Update.*

_____. 2023a. Nationally Determined Contributions Under the Paris Agreement.

_____. 2023b. 2023 NDC Synthesis Report.

_____. n.d. Long-Term Strategies Portal (accessed 26 August 2024).

_____. n.d. Nationally Determined Contributions Registry (accessed 26 August 2024).

_____. n.d. Reporting Requirements.

WEF (World Economic Forum). 2022. China Will Aim to Plant and Conserve 70 Billion Trees by 2030 as Part of the Global Tree Movement. Press Release. 25 May.

WIPO (World Intellectual Property Organization). 2023. *GII 2023 at a Glance.*

World Resources Institute. Climate Watch (accessed 23 August 2024).

ZCA (Zero Carbon Analytics). 2023. *A Driving Force: Asia's Energy Transition.*

Chapter 2

Impacts and Costs of Climate Inaction for Asia and the Pacific

2.1   A Climate in Crisis

2.1.1   Overview

Unmitigated climate change will bring grave consequences for Asia and the Pacific. This chapter provides an overview of the impacts and vulnerabilities that the Asia and Pacific region can face from climate change, drawing on relevant research. It begins by reviewing the global context of climate change and potential trajectories of warming under climate inaction. It then focuses on what warming can mean more specifically for the Asia and Pacific region, biophysically and for exposed assets. Effects on vulnerable natural-resource-based sectors, social systems, and populations are subsequently explored. Finally, consequences of sea-level rise, increased river-based flooding, higher energy demand, and effects on the productivity of labor, agriculture, fisheries, and forestry are considered in an economy-wide model offering new estimates of potential economic losses.

2.1.2   Context

Climate change is no longer an issue about the future. A substantial amount of climate change has already occurred, and the effects will continue to increase even if all greenhouse gas (GHG) emissions were to stop immediately. Global warming is a function of the stock of GHGs in the atmosphere. That stock takes time to respond to changes in the flow of emissions. For this reason, addressing most impacts of climate change in the coming decades depends on adaptation, even if mitigation is critical in the longer term.

Climate change is accelerating. As of mid-2024, the world's atmospheric concentration of carbon dioxide (CO_2) had reached 422 parts per million, which is 50% higher than the preindustrial level of around 280. Two-thirds of the increase has taken place since 1970 and the rate of increase has only accelerated over time. Since the middle of the 18th century, the world has emitted 1.5 trillion tons of CO_2. The amount of CO_2 that could be emitted in the future is a small fraction of this if climate goals under the Paris Agreement on Climate Change are to be met—even as the world is at record levels of population and economic development (IPCC 2022a). As indicated in Figure 2.1, 2023 was already 1.46°C above preindustrial averages. The rate of temperature increase is also accelerating, with warming increasing per year since 1982 three times faster than prior to 1982 (NCEI 2024; WMO 2024).

Figure 2.1: Annual Global Temperature Differences Compared to 1850–1900 Averages

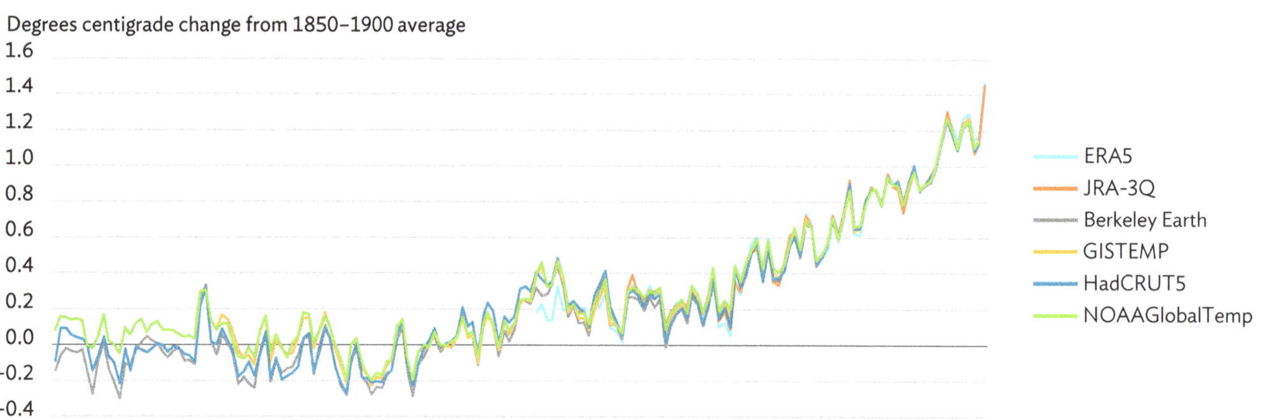

Note: The figure compares global mean temperatures using six data sources that reanalyze historical meteorological data.
Source: Copernicus Climate Change Service (accessed 11 October 2024).

This chapter was written by a team consisting of ADB staff Alessio Giardino, Martino Pelli, and David A. Raitzer (lead author), with consultants Francesco Bosello, Lorenza Campagnolo, and Gabriele Mansi.

Many biophysical impacts for the coming decades are locked in. Over longer time frames, the impacts of climate change are conditioned by whether the world acts in line with Paris Agreement goals of limiting warming to well below 2°C above preindustrial levels and to pursue efforts to limit it to 1.5°C. However, even if strong action were taken, the climate will continue to warm for decades. So far, commitments put forward by individual economies are far from either Paris Agreement goal. If the implemented sector and economic policies are continued, mean warming is expected to reach 3°C above preindustrial levels (ADB 2023; IPCC 2022a). Additional climate measures needed to fulfill nationally determined contributions to the Paris Agreement only slightly change that warming to 2.4°C (ADB 2023). These numbers are based on assumed full implementation of stated climate policies, but there are also implementation gaps that may increase emissions relative to those estimates (Chen and Huan 2019). A warming level of 2.4°C would take the world to a climate system not seen for millions of years.

Warming could happen much faster than implemented policies scenarios depict. As of 2024, emissions have not deviated substantially from the high end emissions scenario of the Intergovernmental Panel on Climate Change (IPCC) termed Shared Socioeconomic Pathway (SSP) 5, Representative Concentration Pathway (RCP) 8.5 or SSP5-8.5.[1] Indeed, GHG concentrations to date are also indistinguishable from those of SSP5-8.5, which reflects a mean estimate of 4.7°C of mean global warming by 2100, compared with the preindustrial era (Schwalm, Glendon, and Duffy 2020). There are additional reasons why the world cannot be complacent that it has avoided a high end emissions scenario. Emissions scenarios to date are based on integrated assessment modeling, which couples economic and emitting sector models with simplified representations of general circulation models (GCMs), the models that attempt to show how the full climate system responds to GHG emissions. Those simplified GCMs lag behind the science of complete GCMs by many years (van Vuuren et al. 2009). As the science has evolved, full GCMs have reflected increased equilibrium climate sensitivity—or more global warming in response to increases in CO_2 concentrations (Bjordal et al. 2020). That increased sensitivity is not yet part of many of the models that conclude that implemented or current policies will lead to 3°C of mean global warming.

The earth's climate and carbon systems have many cycles that are not fully understood. Climate tipping points can trigger feedback effects leading to additional emissions that do not yet figure in the main scenarios of GCMs, much less in integrated assessment models. For example, most solar radiation ends up absorbed in the world's oceans. As a result, global climate change is rapidly warming the world's waters (Venegas, Acevedo, and Treml 2023). This oceanic warming contributes to loss of polar ice cover, which reduces solar reflectance and increases radiation absorbance. At the same time, oceans absorb 30% of CO_2 emissions, but the absorption of CO_2 is falling as a result of climate change (SOCAT 2024).

There are signs the world is reaching tipping points that can change emissions sinks into emissions sources. Tropical forests have historically been major carbon sinks. However, increasing forest fires are releasing the stocks of CO_2 that forests have stored (You 2023), thus turning them into emissions sources. By one estimate, in 2023, global sequestration by forests, plants, and soil fell by 78% compared with the 2010–2022 average, as land sequestration has been negatively affected by warming (Ke et al. 2024). Peatlands, both in boreal areas and in the tropics, store vast amounts of CO_2, but warmer temperatures, drying, and exposure to fire are increasingly releasing these carbon reserves into the atmosphere (Wijedasa et al. 2018). Emissions of methane, a potent GHG, from natural ecosystems have made unprecedented jumps since 2020 (Zhang et al. 2023). In 2024, Antarctic temperatures were at unprecedented highs, threatening vast amounts of high-reflectance sea ice (Colucci 2022). If natural emissions sinks continue to transform to emissions sources, high levels of warming may become impossible to avoid.

---

[1]   This chapter uses SSP5-8.5 ("fossil-fueled development") for reference in the numbers presented, unless otherwise noted. While progress in the development and deployment of renewables may suggest lower radiative forcings under a scenario of no climate policy strengthening, there are considerable climate uncertainties that could lead this to be a plausible level of radiative forcing, even if emissions are substantially lower than integrated assessment models depict for this pathway. While it is a high radiative-forcing scenario, it also does reflect rapid economic growth, which is consistent with the economic development plans of many countries in Asia and the Pacific and dampens economic impacts.

Action to reduce emissions may, paradoxically, lead to more near-term warming. Combustion of fossil fuels produces aerosol air pollutants as well as GHGs. What was not previously understood is that those air pollutants actually reduce the amount of solar radiation absorbed and thereby counteract the levels of warming expected from increases in GHGs. The effect is substantial, as warming without those pollutants could be 30% higher (IPCC 2022a). As the world adopts cleaner energy sources, warming in the near term may actually be intensified relative to existing GCM projections.

2.2   Climate Futures for Asia and the Pacific

Climate change is already having tangible effects. The 10 years across 2014–2023 were the 10 hottest years on record and February 2024 was the hottest month ever reported (C3S 2024). While the Paris Agreement relates to the long-term global average, not an individual year, global average annual temperatures did exceed 1.5°C of warming in the year to January 2024. With this level of warming, Asia and the Pacific experienced unprecedented heat waves during 2024. Large portions of the region experienced many days above 40°C and a number of countries set heat records.

Climate change means a profound change in living conditions for people in Asia and the Pacific. Global mean warming of 4.7°C under a high end emissions scenario will lead to changes in local and seasonal temperatures that can be much higher—since changes vary according to location and season—and temperature rises over land will be greater than over the sea. Patterns in the median GCM projections under the Coupled Model Intercomparison Project Phase 6 (CMIP6), (which informed the IPCC's 6th Assessment Report) for subregions in Asia and the Pacific, as defined by the IPCC, confirms this phenomenon. With 4.7°C of mean global warming in 2100, West and Central Asia[2] would have mean temperature increases of 8°C and East Asia would have mean temperatures that increase by 7°C, whereas Southeast Asia's temperatures would increase by less than 5°C (CICA 2024).

The most dramatic changes will be in heat waves. Historically, each year, Southeast Asia has had 20 days or fewer when the regional temperature reached more than 35°C. Under 4.7°C of mean global warming, such days would exceed half a year by 2100 (Figure 2.2). South Asia would have days exceeding 35°C rise from around 100 to about 200 annually by 2100, East Asia from around 7 to nearly 50, and West and Central Asia from 70 to around 130 (CICA 2024).

---

[2]   This chapter uses regional definitions that differ from the rest of the report. The cited figures define the regional compositions mentioned in the text. Generally, biophysical projections from CMIP6 follows IPCC regional definitions, whereas modeling of impacts breaks out major countries and subregions of developing Asia and the Pacific.

Figure 2.2: Projected Changes in Days Above 35°C Annually Under a High End Emissions Climate Scenario

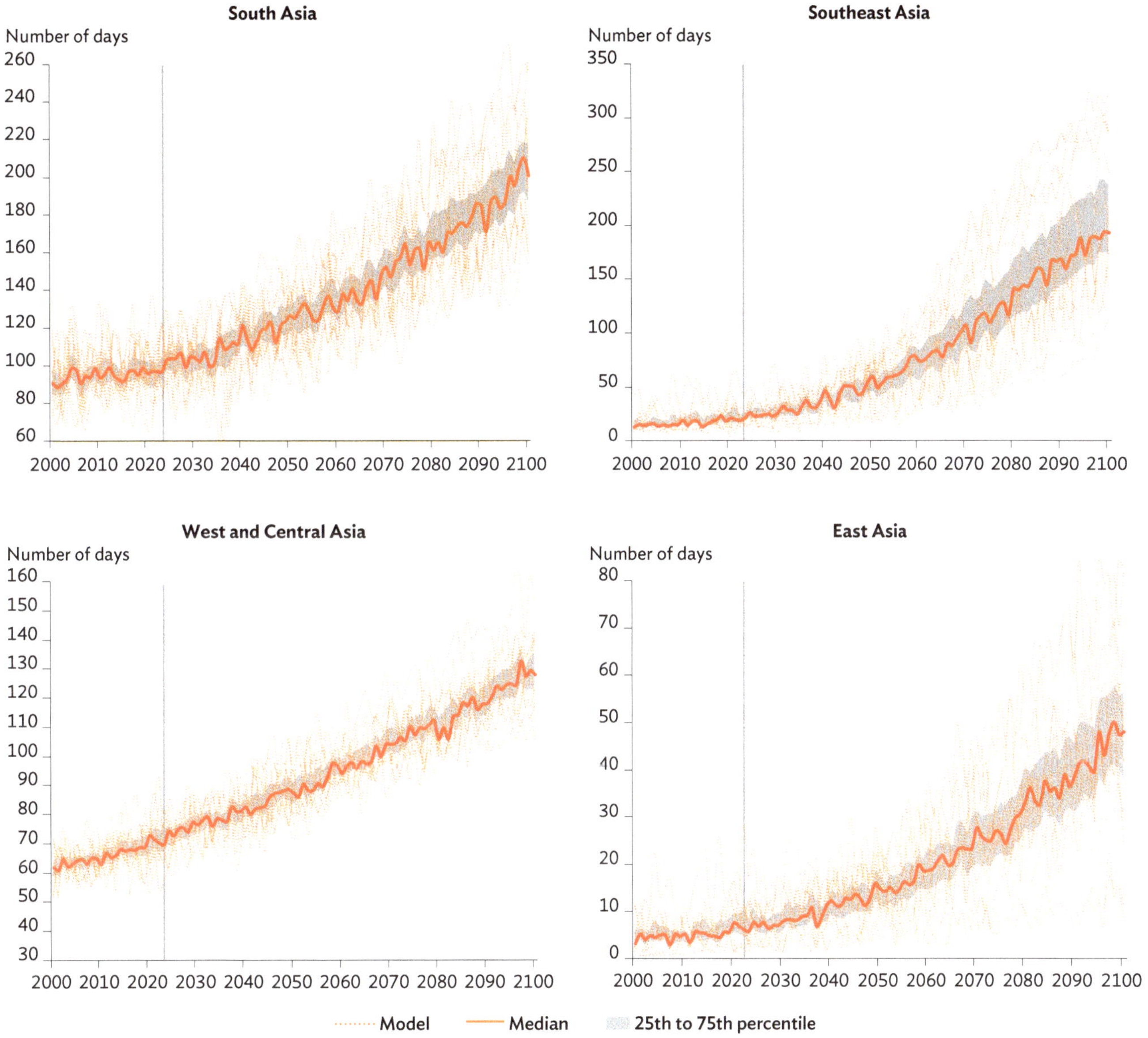

Notes: Light lines depict individual general circulation model projections, while dark lines are medians of the projections under the Coupled Model Intercomparison Project Phase 6. Subregions are as defined by the Intergovernmental Panel on Climate Change, rather than being those of the Asian Development Bank. The scenario depicted is SSP5-8.5.
Source: Copernicus Interactive Climate Atlas (accessed 10 June 2024).

Precipitation patterns are expected to shift under climate change. Most of the Asia and Pacific region is expected to be wetter in the future, as warmer air can hold more moisture and increased temperatures increase oceanic evaporation. However, this overall trend masks substantial variations by locality. In South Asia, precipitation would increase by around 50% by 2100 under 4.7°C of mean global warming and, in East Asia, it would increase by around 25%, whereas Southeast Asia could expect a mixture of increases and decreases. Rainfall will become more extreme and concentrated, as the maximum daily precipitation would rise by about 50% in the Pacific, 40% in South Asia, and 30% in East Asia and Southeast Asia (CICA 2024).

Along with more concentrated rainfall will come more drought events. Under warmer temperatures, more precipitation will be countered by increased evapotranspiration (the combined processes that move water from the earth's surface into the atmosphere, covering both water evaporation from the soil and plant transpiration). This evapotranspiration is expected to increase even more than precipitation. A key measure of drought that assesses the deviation from the long-term water balance (i.e., precipitation minus evapotranspiration) is known as the Standardized Precipitation-Evapotranspiration Index (SPEI), which is taken over 6 months. This index is expected to deviate 5–7 times below the historic mean, under 4.7°C of mean global warming (Figure 2.3). The change is driven by the fact that radiation and temperature are key determinants of evapotranspiration, so that temperature increases lead to evaporative effects that outweigh precipitation increases (CICA 2024).

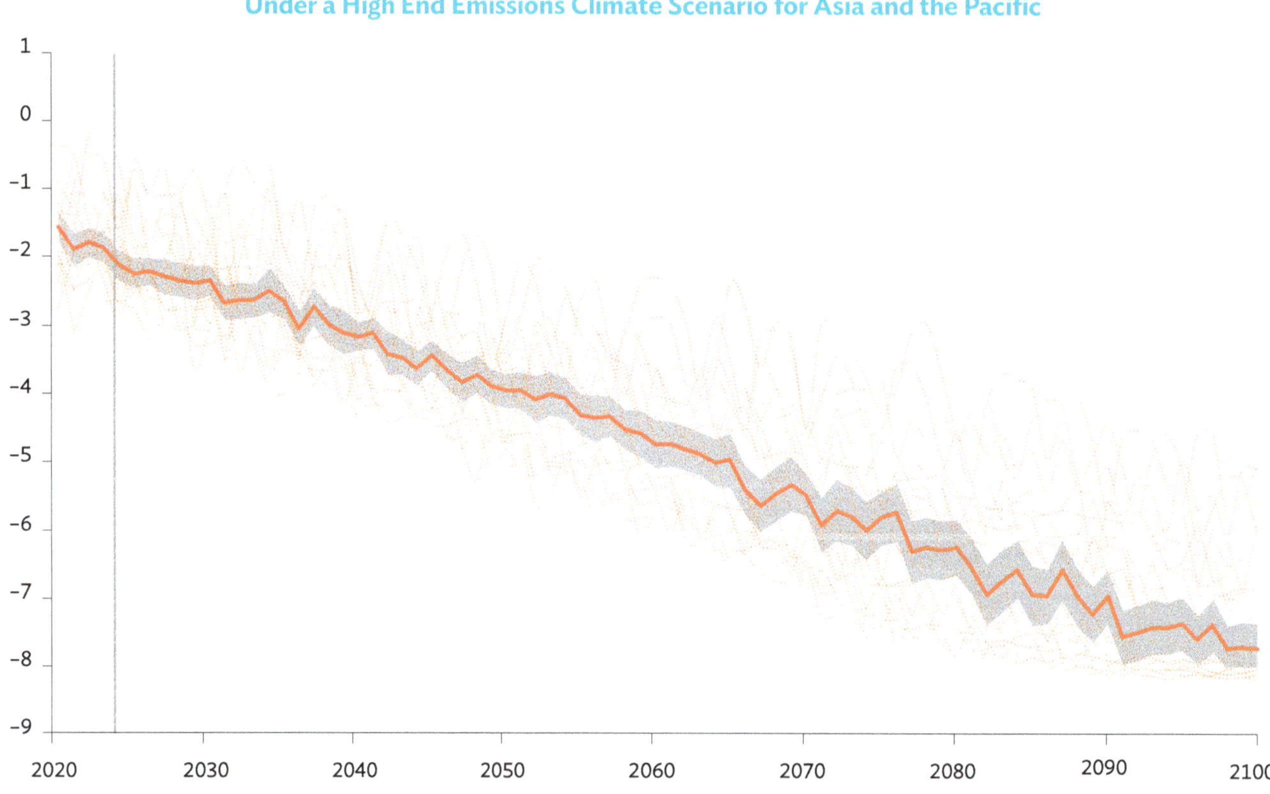

Figure 2.3: Projected Changes in the 6-Month Standardized Precipitation-Evapotranspiration Index Under a High End Emissions Climate Scenario for Asia and the Pacific

Notes: The figure shows the 6-month Standardized Precipitation-Evapotranspiration Index using results taken from the Coupled Model Intercomparison Project. Light lines depict individual general circulation model projections, while dark lines are medians of the projections under the Coupled Model Intercomparison Project Phase 6. The scenario depicted is SSP5-8.5.
Source: Copernicus Interactive Climate Atlas (accessed 12 June 2024).

Extreme storm events across Asia and the Pacific are expected to increase in severity. From 1979 to 2016, cyclones in East Asia and Southeast Asia had a duration that grew by 2–9 hours after making landfall, and they traveled around 100 kilometers further inland (Chen et al. 2021). Under a high end emissions climate scenario, continuation of this pattern would lead to a doubling of the destructive power of cyclones in Asia and the Pacific by 2100, which is much higher than the change expected globally.

Relative sea-level rise in parts of Asia and the Pacific is already twice that of the global average. Sea level is modeled to rise by about 0.8 meters globally by 2100 under a high end emissions scenario, although circulation patterns will introduce local variability (IPCC 2022b). Observed sea-level gauges show rates of relative sea-level rise in certain areas of the Asia and Pacific region that are about double the global average, and internal climate variability is expected to amplify climate change effects in the region to well above those experienced globally (Becker, Karpytchev, and Hu 2023). Arctic and Antarctic instability create additional uncertainty about the degree of sea-level increase, as accelerated melt of polar ice caps can increase this further. Under scenarios of sea ice instability, around 300 million people in Asia and the Pacific live in areas threatened by coastal inundation (Kulp and Strauss 2019). A combination of the increase in mean sea levels as well as amplification in the effect of storm surges and extreme waves will drive substantial inundation risk. Sea-level rise will also lead to salinity intrusion into coastal lands and groundwater (Vousdoukas et al. 2018).

Models reveal trillions of dollars in capital damage each year by sea-level rise, storm surges, wave formation, and coastal flooding in Asia and the Pacific by 2070. Implications of sea-level rise have been computed using the Dynamic Interactive Vulnerability Assessment (DIVA) model globally. DIVA is an integrated modeling framework for coastal systems that assesses the biophysical and socioeconomic consequences of both sea-level rise and socioeconomic development. It takes into account coastal erosion (both direct and indirect), coastal flooding, wetland change, and salinity intrusion into deltas and estuaries. The DIVA model considers three direct climate change impacts: annual land loss, affected population, and asset loss, but it excludes some protection measures and dynamic flooding. To address these limitations, estimates of coastal flooding for the Asia and Pacific region are carried out based on the Large-Scale Integrated Sea-Level and Coastal Assessment Tool (LISCOAST) (Vousdoukas et al. 2018). The modeling framework includes improved data on digital elevation and flood protection levels, use of dynamic flood models to estimate the impacts of extreme events, and incorporation of recent gridded exposure datasets (Giardino and Vousdoukas 2024). Combining LISCOAST for covered countries and subregions with DIVA representation of other countries and subregions leads to the annual damages and affected populations under a high end emissions climate scenario, as shown in Figure 2.4 and Figure 2.5. The largest long-term damages are in the largest economies, while affected populations are concentrated in the People's Republic of China (PRC), India, Bangladesh and Viet Nam. Affected populations triple by 2050 and reach over 50 million annually by 2070. Damages rise by many times to reach a median value of $3 trillion annually by 2070 (considering increases in asset values under economic growth).

Figure 2.4: Annual Damage from Sea-Level Rise Under a High End Emissions Climate Scenario

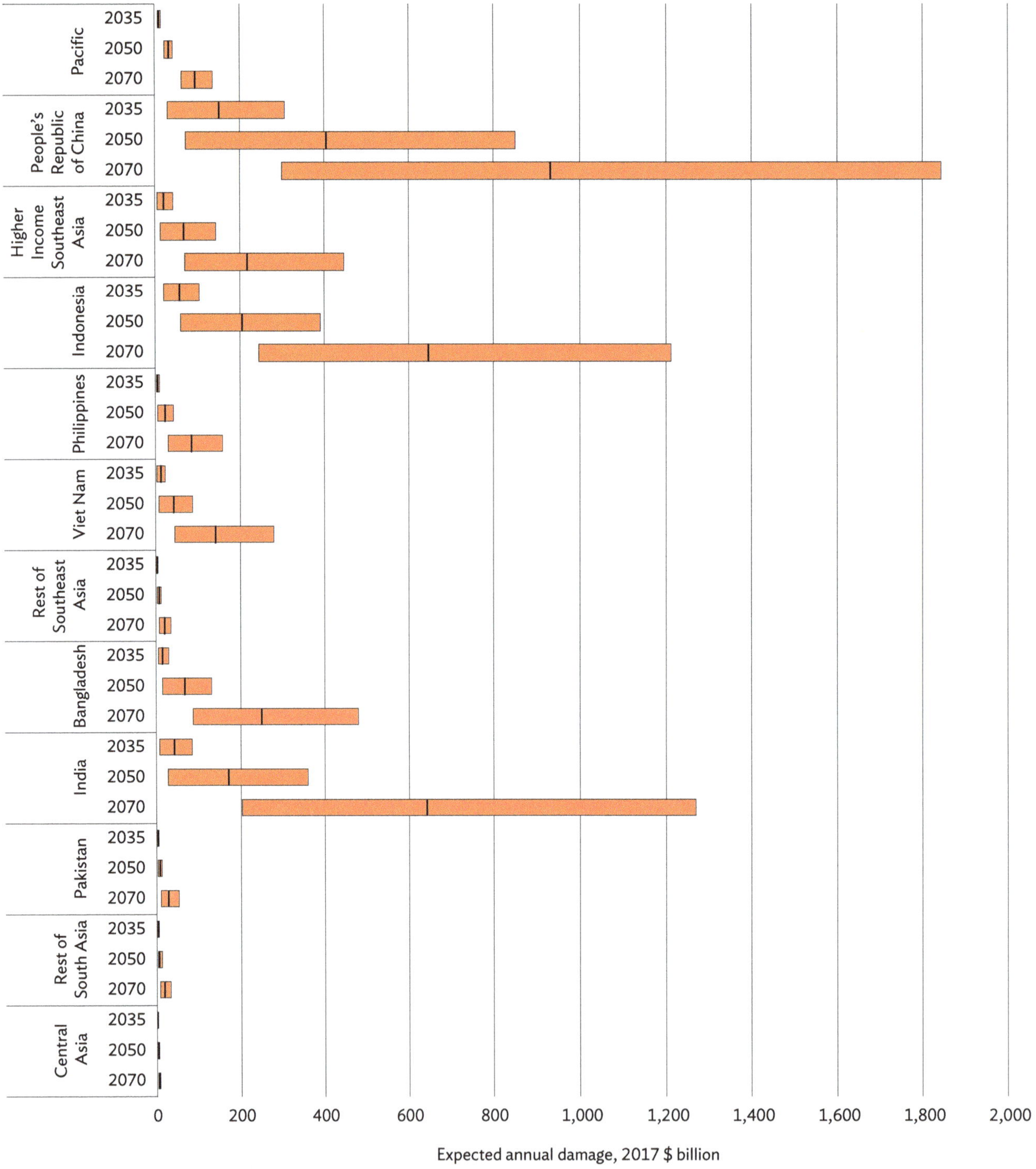

Notes: Bars depict a range of estimates from sea ice melt assumptions, with the central line being the median estimate. The scenario depicted is SSP5-8.5. Central Asia comprises Armenia, Azerbaijan, Georgia, Kazakhstan, the Kyrgyz Republic, Mongolia, Tajikistan, Turkmenistan, and Uzbekistan. Higher Income Southeast Asia comprises Brunei Darussalam, Malaysia, Singapore, and Thailand. The Pacific comprises the Cook Islands, the Federated States of Micronesia, Fiji, Kiribati, the Marshall Islands, Nauru, Niue, Palau, Papua New Guinea, Samoa, Solomon Islands, Tonga, Tuvalu, and Vanuatu. Rest of South Asia comprises Afghanistan, Bhutan, Maldives, Nepal, and Sri Lanka. Rest of Southeast Asia comprises Cambodia, the Lao People's Democratic Republic, Myanmar, and Timor-Leste.
Source: Campagnolo, L., G. Mansi, F. Bosello, and D. A. Raitzer. 2024. Quantifying the Economic Costs of Climate Change Inaction for the Asia-Pacific Region. Background paper for the *Asia–Pacific Climate Report 2024: Catalyzing Finance and Policy Solutions*. Asian Development Bank (results drawn from modeling reported in Giardino, A. and M. Vousdoukas. 2024. Rising Seas: Building Resilience Against Coastal Flooding in Asia and the Pacific. Development Asia. 7 August).

Figure 2.5: Annual Population Affected by Sea-Level Rise Under a High End Emissions Climate Scenario

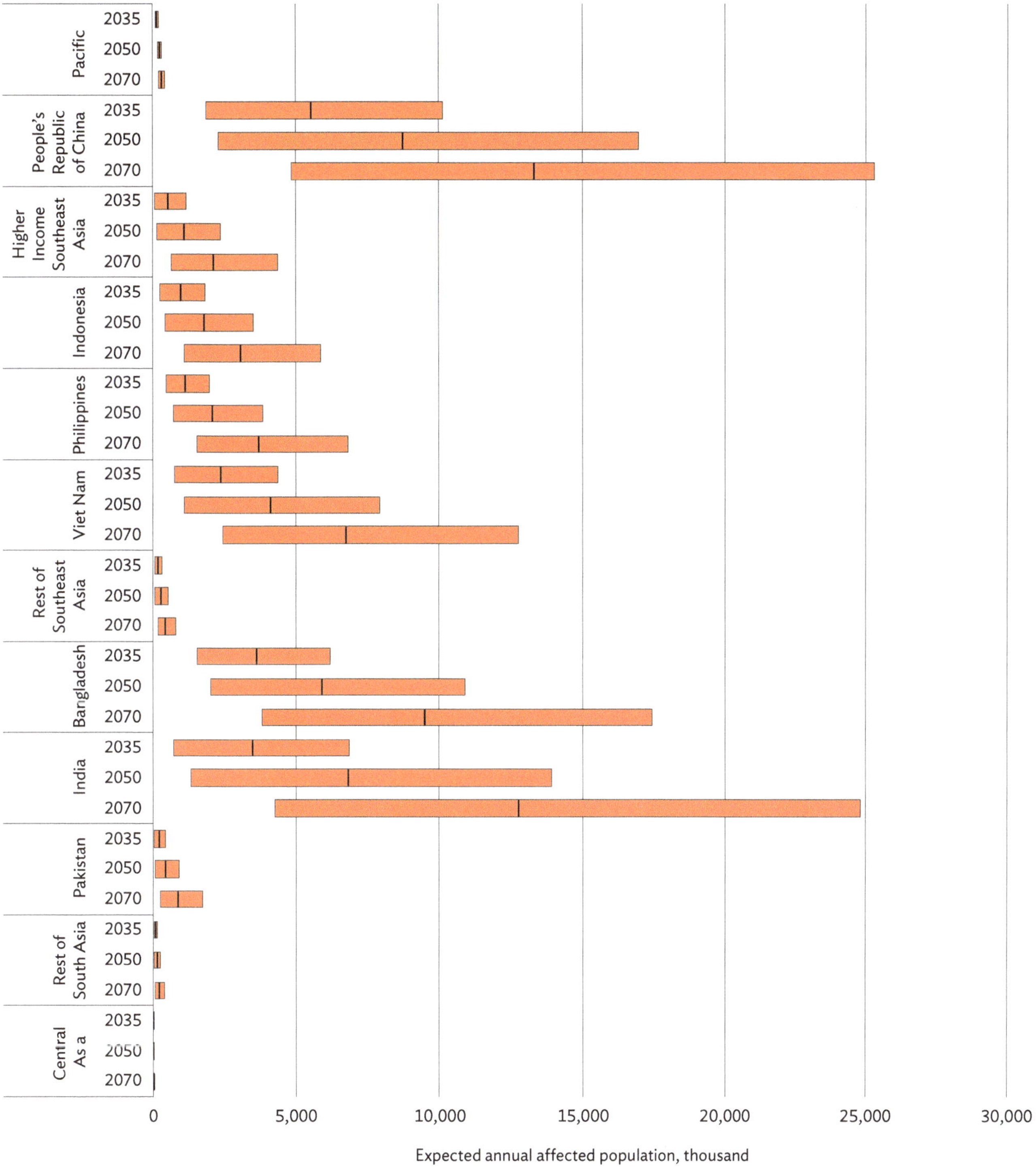

Notes: Bars depict a range of estimates from different sea ice melt assumptions, with the central line being the median estimate. The scenario depicted is SSP5-8.5. Central Asia comprises Armenia, Azerbaijan, Georgia, Kazakhstan, the Kyrgyz Republic, Mongolia, Tajikistan, Turkmenistan, and Uzbekistan. Higher Income Southeast Asia comprises Brunei Darussalam, Malaysia, Singapore, and Thailand. The Pacific comprises the Cook Islands, the Federated States of Micronesia, Fiji, Kiribati, the Marshall Islands, Nauru, Niue, Palau, Papua New Guinea, Samoa, Solomon Islands, Tonga, Tuvalu, and Vanuatu. Rest of South Asia comprises Afghanistan, Bhutan, Maldives, Nepal, and Sri Lanka. Rest of Southeast Asia comprises Cambodia, the Lao People's Democratic Republic, Myanmar, and Timor-Leste.
Source: Campagnolo, L., G. Mansi, F. Bosello, and D. A. Raitzer. 2024. Quantifying the Economic Costs of Climate Change Inaction for the Asia-Pacific Region. Background paper for the *Asia–Pacific Climate Report 2024: Catalyzing Finance and Policy Solutions.* Asian Development Bank (results drawn from modeling reported in Giardino, A. and M. Vousdoukas. 2024. Rising Seas: Building Resilience Against Coastal Flooding in Asia and the Pacific. Development Asia. 7 August).

Intensified and more variable rainfall, along with more extreme storms, will bring increasing landslides and floods. This will be most pronounced in mountainous and steeply sloped areas, such as the border area of India and the PRC, where landslides may increase by 30%–70% under 4.7°C of mean global warming (Kirschbaum et al. 2020). These outcomes will be further worsened by reductions in slope-stabilizing forest cover, as forests unable to cope with new climate regimes suffer dieback.

Flooding will increase, both from overflow of rivers and from intense rainfall leading to water stagnation. IPCC (2022b) concludes that, in several major Asian river basins, the runoff is projected to increase substantially, with estimated increases of 16% for the Brahmaputra, 33% for the Ganges, and 40% for the Meghna. The report also finds that the 50-year return floods in these rivers are likely to increase (IPCC 2022b). This flooding will be particularly problematic in coastal delta megacities, where it can be compounded by the effects of sea-level rise. In addition to flooding from overflow of rivers, increasingly intense rainfall will lead to increased frequency and duration of inundation, as drainage systems struggle to convey accumulated precipitation.

Leading models show trillions of dollars of annual capital damage from riverine flooding in Asia and the Pacific by 2070. Riverine flood impacts have been computed using results from Global Flood Risk with Image Scenarios (GLOFRIS) modeling (Ward et al. 2017; Winsemius et al. 2017), as processed by Standardi et al. (2023). The GLOFRIS model analyzes direct flood damage as a result of riverine and coastal floods under a high end emissions climate scenario. GLOFRIS provides three scenarios that consider average, minimum, and maximum expected damage considering the results of five GCMs. Expected annual damage in the context of economic growth reaches $1.3 trillion per year by 2070, with over 110 million people affected annually. India is found to have the most affected population and damage costs, with residential losses dominant (Figure 2.6 and Figure 2.7).

Figure 2.6: Annual Additional Flood Damage Under a High End Emissions Climate Scenario

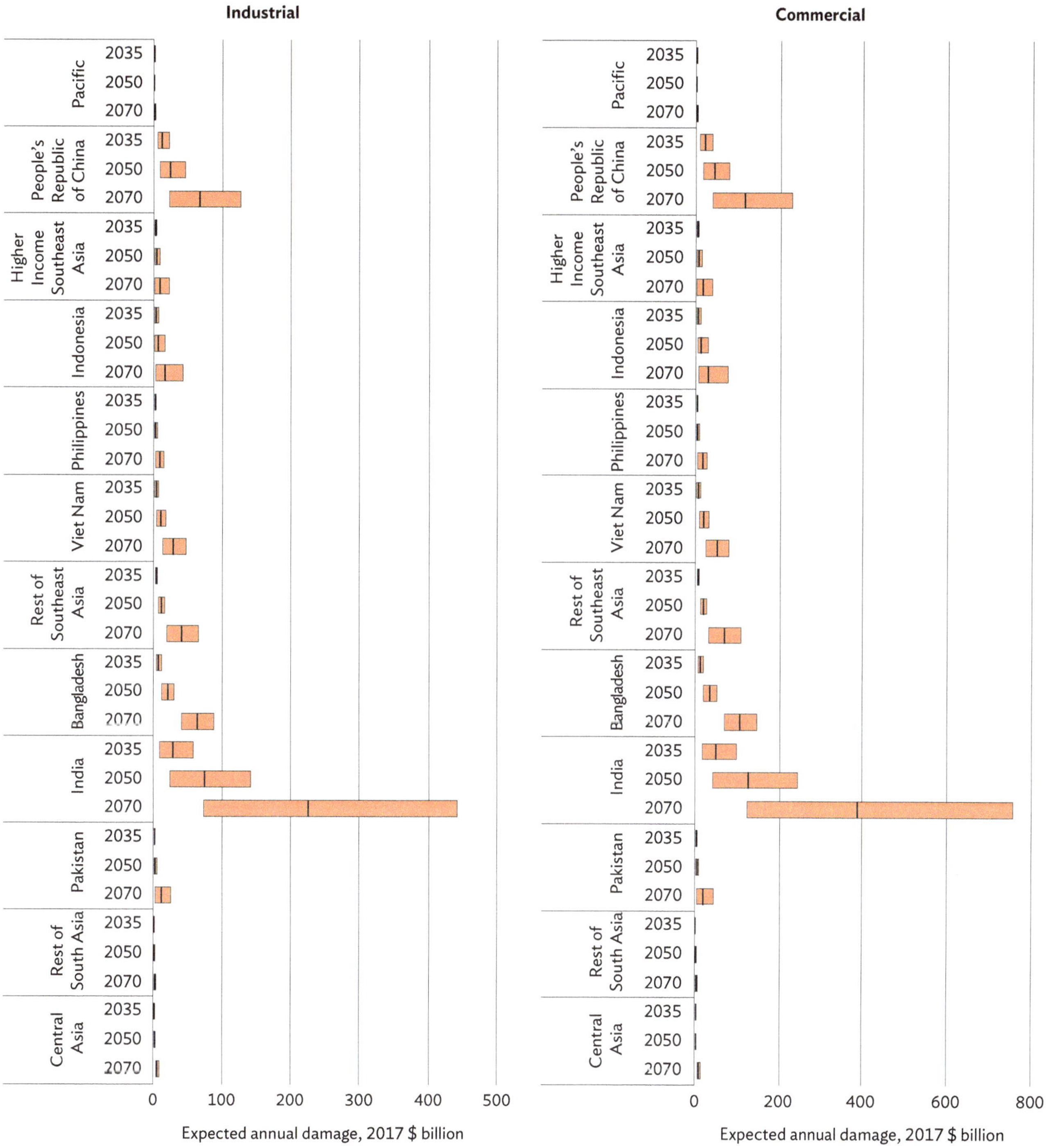

Notes: Bars depict a range of estimates from different general circulation models, with the central line being the median estimate. The scenario depicted is SSP5-8.5. Central Asia comprises Armenia, Azerbaijan, Georgia, Kazakhstan, the Kyrgyz Republic, Mongolia, Tajikistan, Turkmenistan, and Uzbekistan. Higher Income Southeast Asia comprises Brunei Darussalam, Malaysia, Singapore, and Thailand. The Pacific comprises the Cook Islands, the Federated States of Micronesia, Fiji, Kiribati, the Marshall Islands, Nauru, Niue, Palau, Papua New Guinea, Samoa, Solomon Islands, Tonga, Tuvalu, and Vanuatu. Rest of South Asia comprises Afghanistan, Bhutan, Maldives, Nepal, and Sri Lanka. Rest of Southeast Asia comprises Cambodia, the Lao People's Democratic Republic, Myanmar, and Timor-Leste.
Source: Campagnolo, L., G. Mansi, F. Bosello, and D. A. Raitzer. 2024. Quantifying the Economic Costs of Climate Change Inaction for the Asia-Pacific Region. Background paper for the *Asia–Pacific Climate Report 2024: Catalyzing Finance and Policy Solutions*. Asian Development Bank (results drawn from the Co-designing the Assessment of Climate Change Costs Project).

Figure 2.7: Annual Additional Population Affected by Flooding Under a High End Emissions Climate Scenario

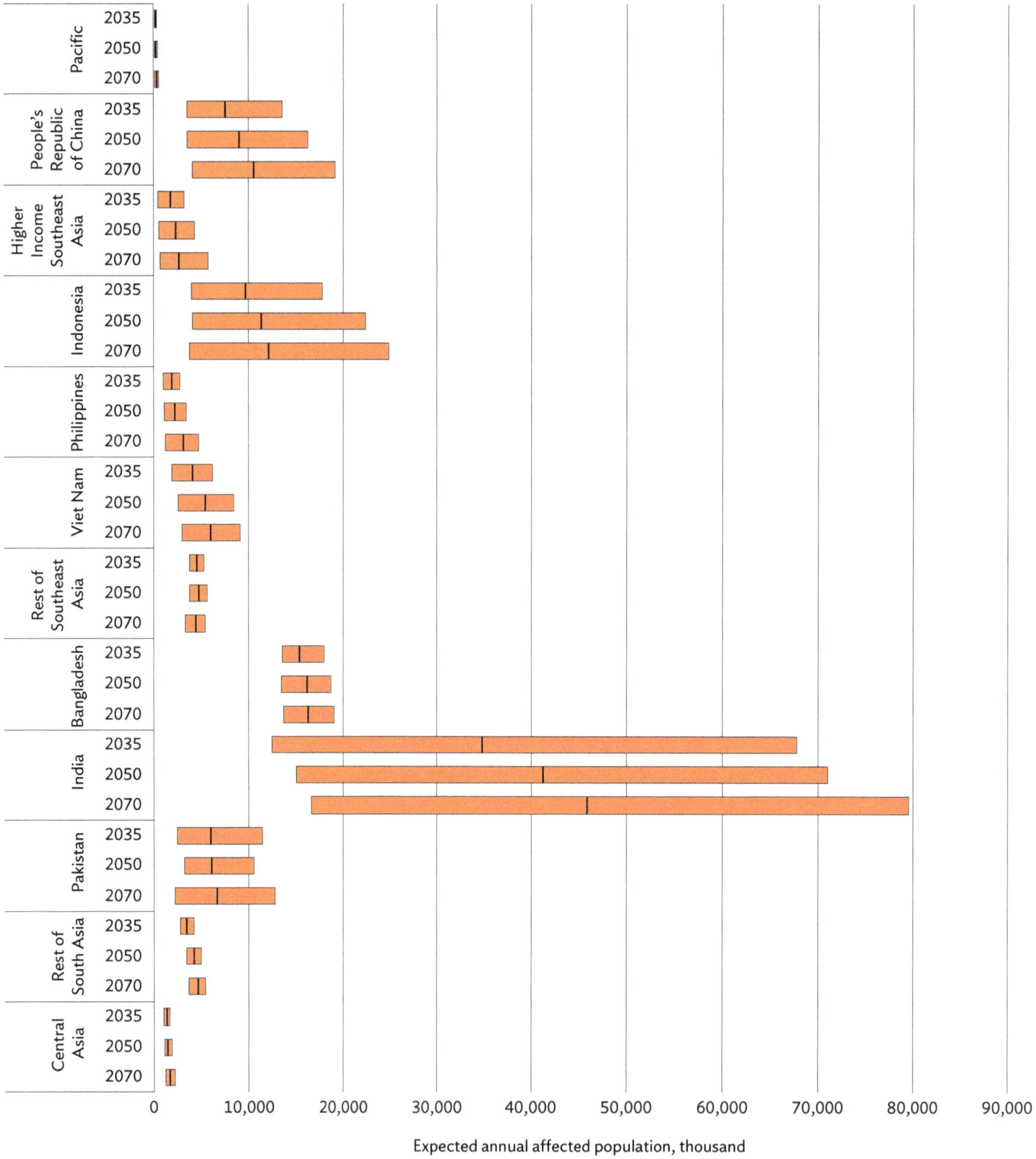

Expected annual affected population, thousand

Notes: Bars depict a range of estimates from different general circulation models, with the central line being the median estimate. The scenario depicted is SSP5-8.5. Central Asia comprises Armenia, Azerbaijan, Georgia, Kazakhstan, the Kyrgyz Republic, Mongolia, Tajikistan, Turkmenistan, and Uzbekistan. Higher Income Southeast Asia comprises Brunei Darussalam, Malaysia, Singapore, and Thailand. The Pacific comprises the Cook Islands, the Federated States of Micronesia, Fiji, Kiribati, the Marshall Islands, Nauru, Niue, Palau, Papua New Guinea, Samoa, Solomon Islands, Tonga, Tuvalu, and Vanuatu. Rest of South Asia comprises Afghanistan, Bhutan, Maldives, Nepal, and Sri Lanka. Rest of Southeast Asia comprises Cambodia, the Lao People's Democratic Republic, Myanmar, and Timor-Leste.

Source: Campagnolo, L., G. Mansi, F. Bosello, and D. A. Raitzer. 2024. Quantifying the Economic Costs of Climate Change Inaction for the Asia-Pacific Region. Background paper for the *Asia–Pacific Climate Report 2024: Catalyzing Finance and Policy Solutions*. Asian Development Bank (results drawn from the Co-designing the Assessment of Climate Change Costs Project).

2.3   Climate Change Impacts on Vulnerable Sectors and Populations

Climate effects on agriculture will disproportionately affect poor and vulnerable communities.
A majority of poor people in the Asia and Pacific region depend on agriculture for livelihoods, and food accounts for the most vulnerable people's consumption, so climate-related impacts on the sector will disproportionately harm the most vulnerable. The region also accounts for around two-thirds of global food production. Agriculture and food systems are inherently vulnerable to the fluctuations and shifts brought on by climate variability and change. This susceptibility stems from the deep dependence of those systems on weather and environmental conditions, which directly affect water availability, water requirements, heat stress, and pest and disease patterns. Agriculture also depends on broader ecosystem services, such as pollination, which are threatened by changes in ecological niches.

Changes will be most pronounced in terms of increasing extremes. Climate change is expected to intensify the El Niño phenomenon, so that losses from heat and wetness extremes will be more frequent and severe (Wilcox et al. 2023). Drought events are anticipated to increase in frequency, with populations exposed to drought multiplying several times under a high end emissions scenario by 2100, especially in key localities such as the Pearl River Basin (Duan et al. 2021). With more severe extremes, it will be more difficult for farmers to optimize inputs and manage risk, increasing impacts. Biophysical effects on crops will also be exacerbated by effects on labor, as agriculture involves many manual operations under conditions in which personal cooling is difficult. In many developing countries of Asia and the Pacific, agricultural operations are already constrained by labor limitations (as former farm laborers have migrated), and climate change will compound the effects of labor scarcity.

Higher CO_2 concentrations can partially counter other climate change impacts on crops. Carbon dioxide fertilization, or the increase in photosynthesis possible under higher CO_2 concentrations, will partially offset losses due to growing conditions that are otherwise less favorable for crops. As a result, recent crop growth modeling exercises that incorporate this effect, along with other climate modeling projections, find relatively small losses to crop yields from climate change globally and in Asia and the Pacific (Ruane et al. 2024). The Inter-Sectoral Impact Model Intercomparison Project (ISIMIP) has used gridded growth models of crop yields (ISIMIP 2024) under different climate scenarios,[3] and has found losses for a high end emissions climate change scenario (SSP5-8.5) with respect to 2020[4] for maize, but more limited losses for the major wheat-, rice-, and soy-producing zones in Asia and the Pacific (Figure 2.8). For soybeans, projected yield under ISIMIP moderately benefits from climate change in the short and medium term (to 2050), but a negative pattern becomes clear in the longer term (to 2070). Overall, rice yield improves under climate change in almost every economy or subregional aggregate, excluding Pakistan and Central Asia, whereas wheat shows strong regional differences.

---

[3]   The following crop models have been selected due to crop coverage and scenario availability: CROVER, EPIC-IIASA, LDNDC, LPJmL, PEPIC, CYGMA1P74; each crop model provides model runs using different global climate models (GCMs): GFDL-ESM4, IPSL-CM6A-LR, MPI-ESM1-2-HR, MRI-ESM2-0, UKESM1-0-LL. The models have been smoothed.

[4]   Crop models do not account for any socioeconomic forcing apart from those observed in the base year; therefore, all the observed yield change can be attributed to the climate forcing.

Figure 2.8: Climate Change Effects on Crop Yields Under a High End Emissions Scenario as Shown by Gridded Crop Growth Modeling

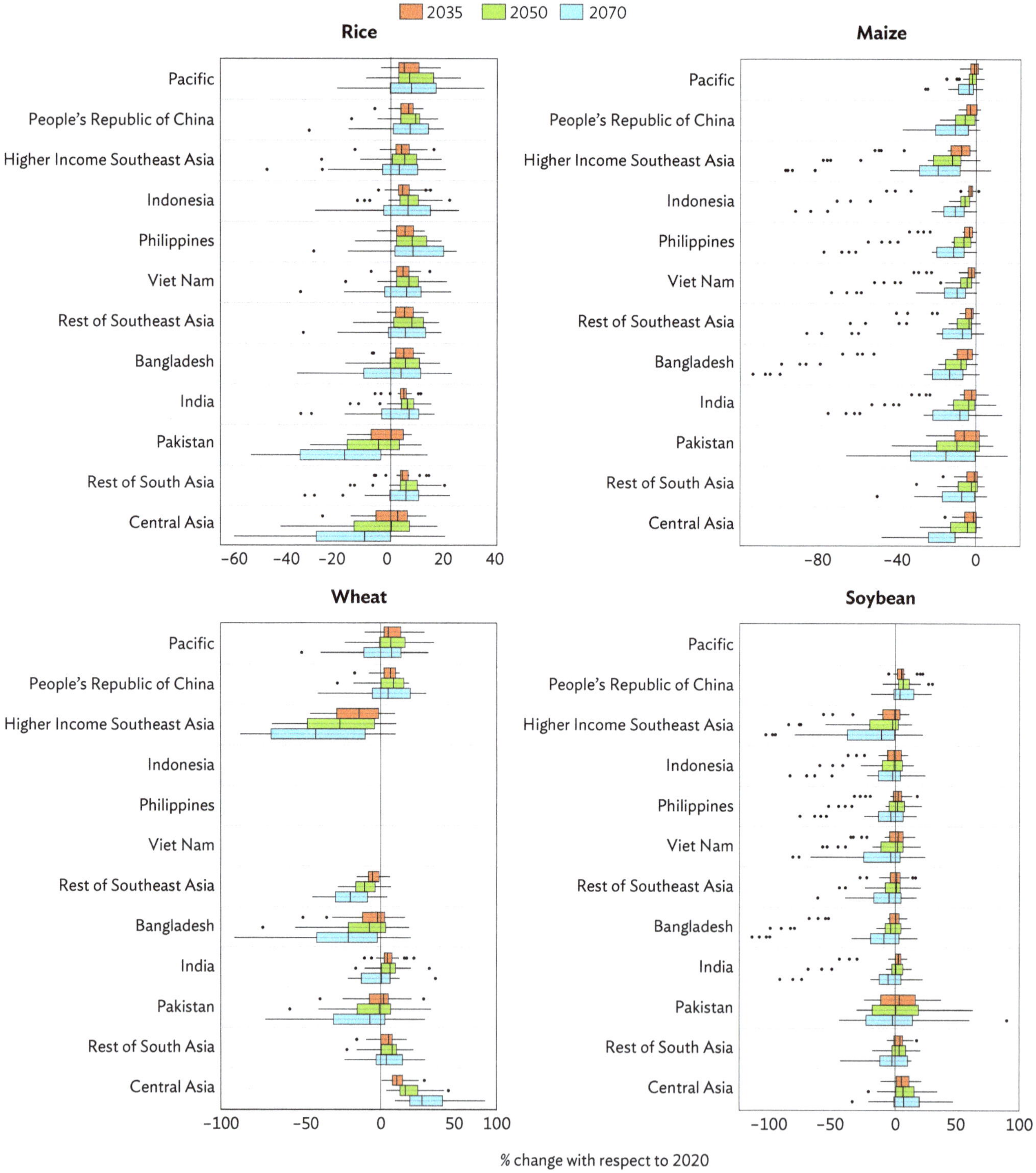

Notes: The scenario depicted is SSP5-8.5. Central Asia comprises Armenia, Azerbaijan, Georgia, Kazakhstan, the Kyrgyz Republic, Mongolia, Tajikistan, Turkmenistan, and Uzbekistan. Higher Income Southeast Asia comprises Brunei Darussalam, Malaysia, Singapore, and Thailand. The Pacific comprises the Cook Islands, the Federated States of Micronesia, Fiji, Kiribati, the Marshall Islands, Nauru, Niue, Palau, Papua New Guinea, Samoa, Solomon Islands, Tonga, Tuvalu, and Vanuatu. Rest of South Asia comprises Afghanistan, Bhutan, Maldives, Nepal, and Sri Lanka. Rest of Southeast Asia comprises Cambodia, the Lao People's Democratic Republic, Myanmar, and Timor-Leste.
Source: Campagnolo, L., G. Mansi, F. Bosello, and D. A. Raitzer. 2024. Quantifying the Economic Costs of Climate Change Inaction for the Asia-Pacific Region. Background paper for *Asia–Pacific Climate Report 2024: Catalyzing Finance and Policy Solutions*. Asian Development Bank (results drawn from the Intersectoral Model Intercomparison Project 3b database accessed 1 March 2024).

Effects of climate change on agriculture via changes to weather extremes may be profound. It is well established that process-based crop growth models poorly replicate impacts of droughts and other extreme events. These models also omit many other important impact channels, such as effects on pollination, pests and diseases, labor, salinity intrusion, or increased competition for energy and water under climate change (Heinicke et al. 2022). Effects of the precipitation-evapotranspiration extremes have been empirically estimated for this chapter and projected for a high end emissions climate scenario (Box 2.1; Raitzer and Drouard 2024). They subsequently have been combined with crop growth modeling results for mean climatic conditions to give a more complete picture of climate change consequences. Figure 2.9 illustrates the combined results by 2100, which indicate that wheat may have yield reductions exceeding 40% in major countries of South Asia and maize may have reductions exceeding 20% in much of Asia, whereas rice is modestly or positively affected.

Figure 2.9: Cumulative Climate Change Impacts on Crop Yields Under a High End Emissions Scenario

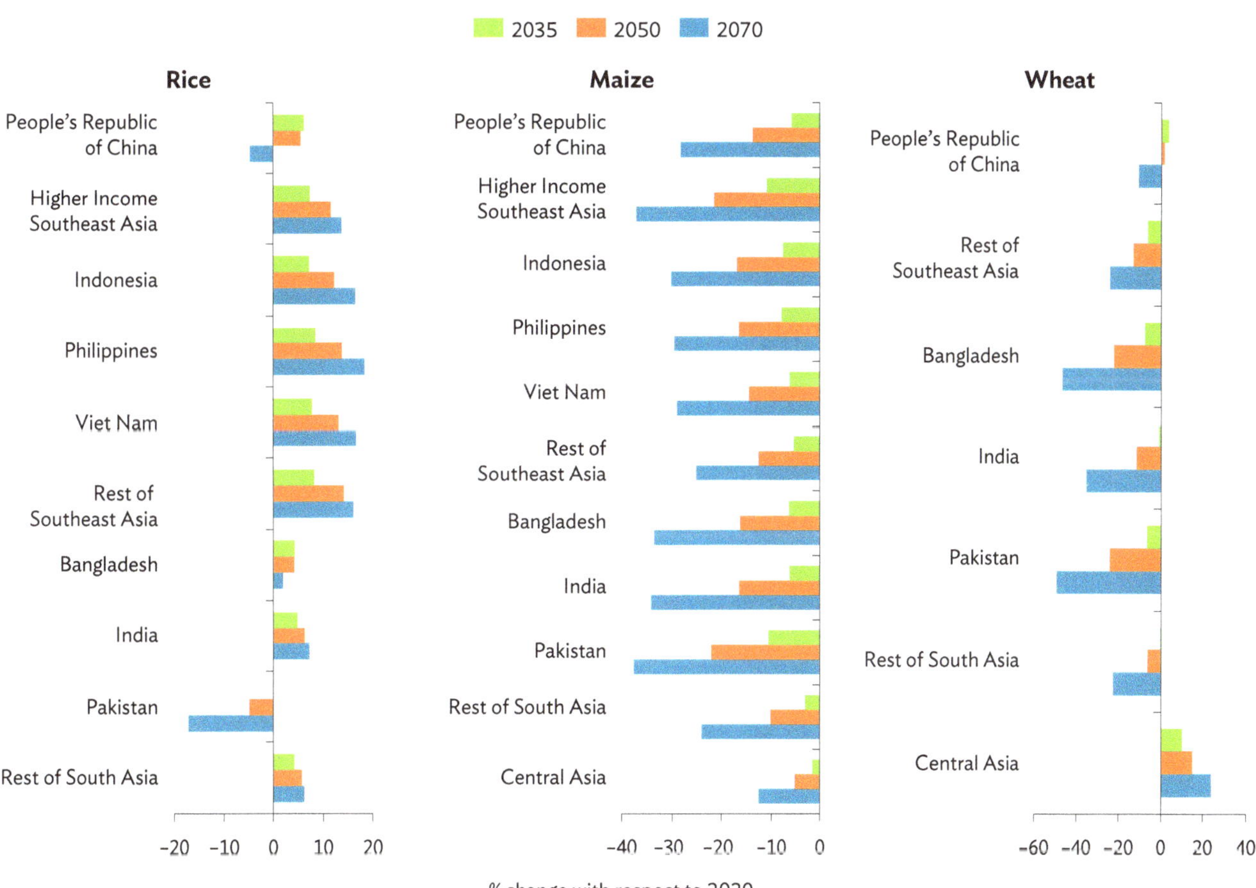

Notes: The analysis in the figure considers both gridded crop growth modeling results and econometric estimates of effects of precipitation evapotranspiration extremes. The scenario depicted is SSP5-8.5. Central Asia comprises Armenia, Azerbaijan, Georgia, Kazakhstan, the Kyrgyz Republic, Mongolia, Tajikistan, Turkmenistan, and Uzbekistan. Higher Income Southeast Asia comprises Brunei Darussalam, Malaysia, Singapore, and Thailand. The Pacific comprises the Cook Islands, the Federated States of Micronesia, Fiji, Kiribati, the Marshall Islands, Nauru, Niue, Palau, Papua New Guinea, Samoa, Solomon Islands, Tonga, Tuvalu, and Vanuatu. Rest of South Asia comprises Afghanistan, Bhutan, Maldives, Nepal, and Sri Lanka. Rest of Southeast Asia comprises Cambodia, the Lao People's Democratic Republic, Myanmar, and Timor-Leste.
Source: Campagnolo, L., G. Mansi, F. Bosello, and D. A. Raitzer. 2024. Quantifying the Economic Costs of Climate Change Inaction for the Asia-Pacific Region. Background paper for the *Asia–Pacific Climate Report 2024: Catalyzing Finance and Policy Solutions*. Asian Development Bank (results drawn from the Intersectoral Model Intercomparison Project 3b database accessed 1 March 2024; and combined with results from Raitzer, D. A. and J. Drouard. 2024. Empirically Estimated Impacts of Climate Change on Global Crop Production via Increasing Precipitation-Evapotranspiration Extremes. Background paper for the *Asia–Pacific Climate Report 2024: Catalyzing Finance and Policy Solutions*. Asian Development Bank).

Box 2.1: Impacts of Climatic Extremes on Crop Yields Under Climate Change

Although agriculture has often been considered as vulnerable to climate change, recent gridded crop growth modeling exercises have found that yields of staple crops will benefit from, or be modestly affected by, rapid global warming. However, those crop growth models also do not fully reflect the impacts of increasing climate extremes.

New analysis by Raitzer and Drouard (2024) uses global remote-sensing-derived yield data, global agro-meteorological reanalysis data, and precipitation-evapotranspiration data derived from weather stations and remote sensing to construct a grid cell panel data set at 0.1 degree resolution for 2003–2015. Regressions that control for grid-cell-specific intercepts and time trends, temperature, rainfall, and cloud cover empirically estimate the relationship between yields and extreme weather for each growing season and crop of rice, wheat, and maize in each subregion of the globe as distinct from small changes in production conditions. Estimated coefficients are applied to projections of precipitation-evapotranspiration extremes under a high end emissions scenario in each grid cell, from an ensemble mean of 25 Coupled Model Intercomparison Phase 6 general circulation models to project losses from climate change. The results confirm large impacts on yields from changes in weather extremes. All crops are found to have substantial future global yield reductions, but reductions are highest for wheat and maize, with losses most pronounced in southern Asia and southern Africa, as detailed in the table below.

Effects of Precipitation-Evapotranspiration Extremes on Average Crop Yields, Under a High End Emissions Climate Scenario
(% change relative to 2003–2015 average)

	Rice Production			Wheat Production			Maize Production		
	2030	2050	2090	2030	2050	2090	2030	2050	2090
Australia and New Zealand				–9.2	–21.4	–49.8			
Caribbean	–4.0	–8.1	–19.2				1.1	3.2	10.6
Central America	–1.3	–4.8	–16.6	–3.7	–8.9	–26.3	–11.0	–21.8	–46.9
Central Asia				–0.6	–1.3	–3.8	–3.5	–8.9	–23.8
Eastern Africa	–3.5	–10.7	–33.4	–0.8	–3.4	–14.0	–4.6	–13.4	–40.5
Eastern Asia	–6.0	–13.2	–32.5	–7.5	–16.4	–37.0	–3.3	–10.3	–33.7
Eastern Europe							–2.6	–4.4	–9.7
Middle Africa	–5.3	–12.2	–28.6				–8.6	–22.0	–54.4
Northern Africa	–5.8	–12.2	–30.0	–11.9	–19.6	–36.6	–2.4	–7.1	–23.5
Northern America	–4.8	–11.1	–30.3	–4.0	–9.1	–23.3	–7.9	–17.6	–43.6
Northern Europe				–4.8	–15.9	–46.4			
South America	–2.9	–5.2	–10.0	–1.4	–5.6	–18.0	–1.9	–7.7	–29.1
South-Eastern Asia	4.3	8.0	17.5				–6.4	–14.5	–34.3
Southern Africa				–10.4	–27.7	–65.0	–1.6	–10.0	–41.0
Southern Asia	–0.2	–0.2	0.3	–6.1	–23.7	–65.2	–4.2	–14.7	–43.6
Southern Europe	0.5	–0.4	–5.1	–2.5	–7.3	–27.1	–4.0	–6.2	–10.8
Western Africa	–1.5	–4.8	–14.7				–4.2	–10.4	–27.2
Western Asia				–13.9	–35.5	–73.4	0.3	1.3	5.8
Western Europe				–2.1	–7.2	–27.5	–2.7	–4.3	–7.7

Note: The scenario depicted is SSP5-8.5.

Reference:

Raitzer, D. A. and J. Drouard. 2024. Empirically Estimated Impacts of Climate Change on Global Crop Production via Increasing Precipitation-Evapotranspiration Extremes. Background paper for the Asia–Pacific Climate Report 2024: Catalyzing Finance and Policy Solutions. Asian Development Bank.

Source: Authors.

Oceans absorb around 70% of the solar radiation trapped by the earth's atmosphere and this absorption is profoundly changing marine ecosystems. Sea temperatures are projected to increase by nearly the same amount as global mean air temperatures, changing the suitability of habitats for flora and fauna and throwing food webs out of balance. Absorption of CO_2 has made oceans more acidic, harming coral reefs. More than 90% of coral reefs in the Pacific are threatened by climate change and/or other anthropometric pressures (Burke et al. 2012). Global fish stocks are already threatened by overfishing, and climate change will compound the problem as declines in the productivity of fisheries will lead to higher prices that incentivize more fishing and greater competition for the remaining stocks.

Leading models show that fish yields will fall sharply under a high end emissions scenario. Climate change impacts on fishing activities have been computed by ISIMIP impact models, Bioeconomic Marine Trophic Size-Spectrum (BOATS) and EcoOcean, providing results for two GCMs. The yield of fishing activities is proxied by the total consumer biomass density (grams per square meter), i.e., the density of all animals, size classes, or trophic groups. With the exception of Bangladesh and Central Asia, all assessed countries and subregions are found to have reduced fish availability, even in the near term. Loss of yield is projected to consistently increase by 2070, with nearly every economy and subregion facing declines of 10% to 25% in fishing sector productivity with respect to 2020 (Figure 2.10).

Figure 2.10: Climate Change Impacts on Fishing Catch Under a High End Emissions Scenario

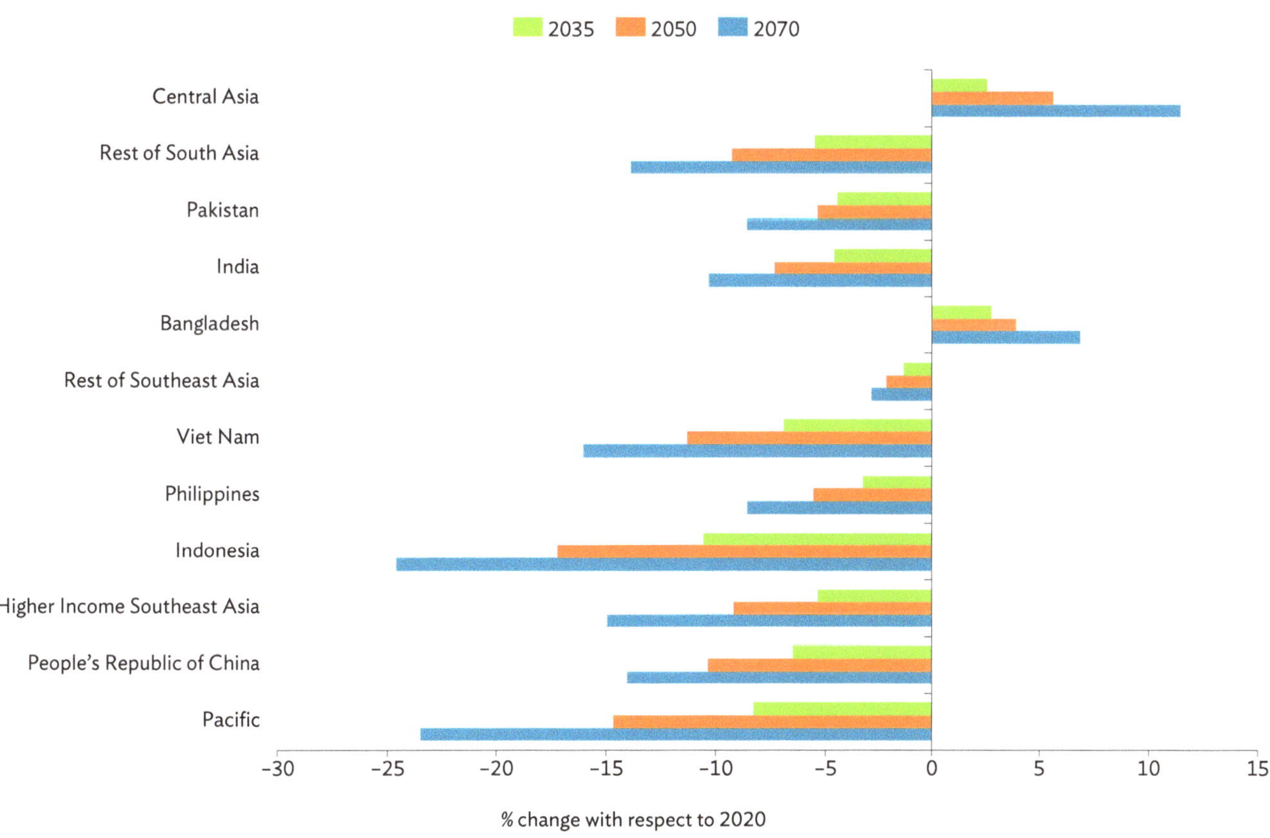

Notes: The scenario depicted is SSP5-8.5. Central Asia comprises Armenia, Azerbaijan, Georgia, Kazakhstan, the Kyrgyz Republic, Mongolia, Tajikistan, Turkmenistan, and Uzbekistan. Higher Income Southeast Asia comprises Brunei Darussalam, Malaysia, Singapore, and Thailand. The Pacific comprises the Cook Islands, the Federated States of Micronesia, Fiji, Kiribati, the Marshall Islands, Nauru, Niue, Palau, Papua New Guinea, Samoa, Solomon Islands, Tonga, Tuvalu, and Vanuatu. Rest of South Asia comprises Afghanistan, Bhutan, Maldives, Nepal, and Sri Lanka. Rest of Southeast Asia comprises Cambodia, the Lao People's Democratic Republic, Myanmar, and Timor-Leste.
Source: Campagnolo, L., G. Mansi, F. Bosello, and D. A. Raitzer. 2024. Quantifying the Economic Costs of Climate Change Inaction for the Asia-Pacific Region. Background paper for the *Asia–Pacific Climate Report 2024: Catalyzing Finance and Policy Solutions*. Asian Development Bank (results drawn from the Intersectoral Model Intercomparison Project 3b database accessed 1 March 2024).

Forests contain slow-growing tree species with limited ability to adapt to rapid shifts in climate regimes within their lifespans. At the same time, shifts in climate enable new pests, parasites, diseases, and exotic competitor species to spread into previously unaffected areas (IPCC 2022b). Increased frequency of dry spells can cause more dry matter to be flammable, increasing forest fire risks. The resulting phenomenon of forest dieback will mean that ecosystem services from natural forest areas will decline, affecting the ability to adapt to other climate changes.

Modeling illustrates substantial losses in forest productivity under climate change. The Global Forest Model or G4M (IIASA 2021) has been used for computing average forest productivity under a high end emissions climate change scenario (SSP5-8.5) with respect to 2020. Figure 2.11 shows the cumulative percentage change in regional forest productivity with respect to 2020, under a high end emissions scenario (SSP5-8.5). A negative pattern is clear for every economy or subregion assessed, and particularly for Viet Nam, India, "the rest of Southeast Asia," and high-income Southeast Asia. Overall, climate shocks on forest productivity have been found as moderate in the short term (to 2030), while larger losses are found for 2070, with a decrease of 10%–30% for most countries or subregions. Only a few localities (the PRC, Pakistan, "the rest of South Asia," and Central Asia) can expect losses below 5% in the long term.

Figure 2.11: Climate Change Impacts on Forest Productivity Under a High End Emissions Scenario

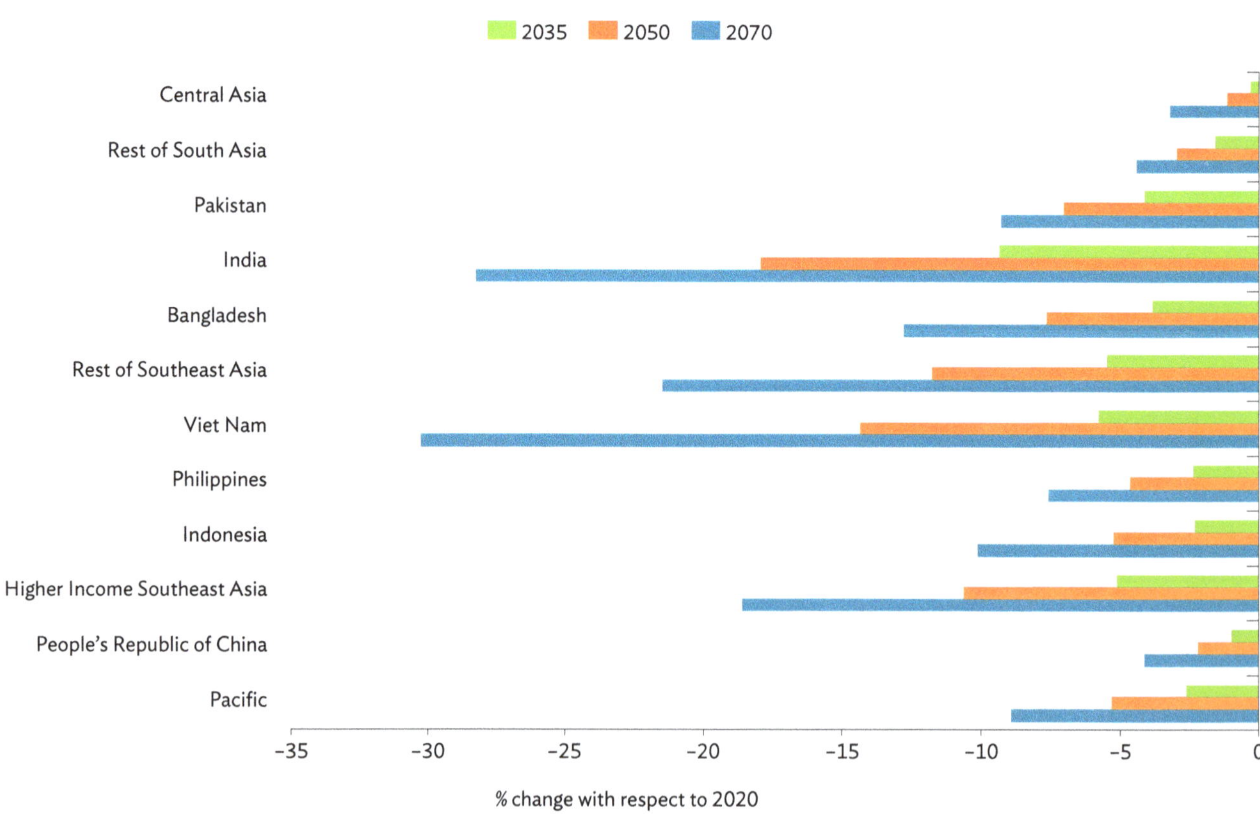

Notes: The scenario depicted is SSP5-8.5. Central Asia comprises Armenia, Azerbaijan, Georgia, Kazakhstan, the Kyrgyz Republic, Mongolia, Tajikistan, Turkmenistan, and Uzbekistan. Higher Income Southeast Asia comprises Brunei Darussalam, Malaysia, Singapore, and Thailand. The Pacific comprises the Cook Islands, the Federated States of Micronesia, Fiji, Kiribati, the Marshall Islands, Nauru, Niue, Palau, Papua New Guinea, Samoa, Solomon Islands, Tonga, Tuvalu, and Vanuatu. Rest of South Asia comprises Afghanistan, Bhutan, Maldives, Nepal, and Sri Lanka. Rest of Southeast Asia comprises Cambodia, the Lao People's Democratic Republic, Myanmar, and Timor-Leste.
Source: Campagnolo, L., G. Mansi, F. Bosello, and D. A. Raitzer. 2024. Quantifying the Economic Costs of Climate Change Inaction for the Asia-Pacific Region. Background paper for the *Asia–Pacific Climate Report 2024: Catalyzing Finance and Policy Solutions.* Asian Development Bank (results drawn from the Co-designing the Assessment of Climate Change Costs Project).

Climate change will also adversely affect social and human capital. A range of empirical studies confirm that violence, criminality, and conflict increase under warmer peak temperatures. Intimate and gender-based violence also increases during heat waves, which will become more common under climate change (Sanz-Barbero et al. 2018; Hu et al. 2017). Learning outcomes worsen when classroom conditions increase beyond optimal temperatures (Park et al. 2020). Increasing storms and extreme weather events also lead to educational disruptions that further depress learning progress (Box 2.2). Many of these impacts will disproportionately affect girls.

Box 2.2: Impacts of Climatic Shocks on Physical and Human Capital in India

Pelli et al. (2023) examine how the capital stock of Indian manufacturing firms responds to storms, utilizing a panel dataset from 1995 to 2006. The paper finds that intense storms (90th percentile) lead to a 5.5% reduction in fixed assets, while average storms reduce fixed assets by 2.2%. Additionally, intense storms decrease sales by 7.8%, and average storms lead to a 3.1% drop in sales. Bantasan et al. (2024), using estimated coefficients together with various survey and census datasets find that intense storms caused significant damage, affecting 0.2% of the total fixed capital (approximately $2.8 billion) and 0.3% of total sales (around $14.5 billion in 2021).

Pelli and Tschopp (2023), focus on the effects of storms on educational outcomes and labor market prospects in India, i.e., impacts on human capital. Using a measure of childhood storm exposure, based on the measure described above, the study reveals that superstorms increase the likelihood of no formal education by 4.6 percentage points and reduce higher educational attainment by up to 20 percentage points, indicating long-term deskilling of affected areas. The reduction in lifetime aggregate earnings, attributed to diminished educational attainment due to a yearly average storm exposure, is estimated at approximately $15.1 billion in 2021 (Bantasan et al. 2024).

References:
Bantasan, D., N. Charette, M. Pelli, and J. Tschopp. 2024. Beneath the Clouds: A Microdata Approach to Understanding Storm Impacts in India. Background paper for the *Asia–Pacific Climate Report 2024: Catalyzing Finance and Policy Solutions*. Asian Development Bank.
Pelli, M., J. Tschopp, N. Bezmaternykh, and K. M. Eklou. 2023. In the Eye of the Storm: Firms and Capital Destruction in India. *Journal of Urban Economics*. 134.
Pelli, M. and J. Tschopp. 2023. Storms, Early Education and Human Capital.

Source: Authors.

Coping with higher temperatures will require more energy for additional cooling. While additional energy requirements in colder climates may be offset by reductions in heating costs, much of developing Asia and the Pacific is in warm climates where there is little heating cost reduction. The Asia and Pacific region includes the countries with the fastest increase in cooling degree days (a measure of demand for cooling) globally. This will drive sharply increased electricity demand. For example, by 2040, it is expected that 30% of electricity use in Southeast Asia will be for cooling, compared with 10% in 2019 (IEA 2019).

Empirical estimates imply strong increases in electricity demand for cooling. Econometric estimates of energy demand elasticity to hot or cold days from De Cian and Sue Wing (2019) have been used by Standardi et al. (2023) to project effects of increased temperatures on demand for electricity, petroleum products, and natural gas for the agriculture, industry, services, and residential sectors. Future energy demand under a high emissions climate scenario for gas and electricity is expected to increase for every economy and subregion assessed with respect to 2020, although the growth of gas demand is moderate (below 20%) even in the long term (to 2070), with the exception of the PRC, which has an increase of more than 80% compared to 2020 (Figure 2.12).

Figure 2.12: Climate Change Impacts on Final Energy Demand Under a High End Emissions Scenario

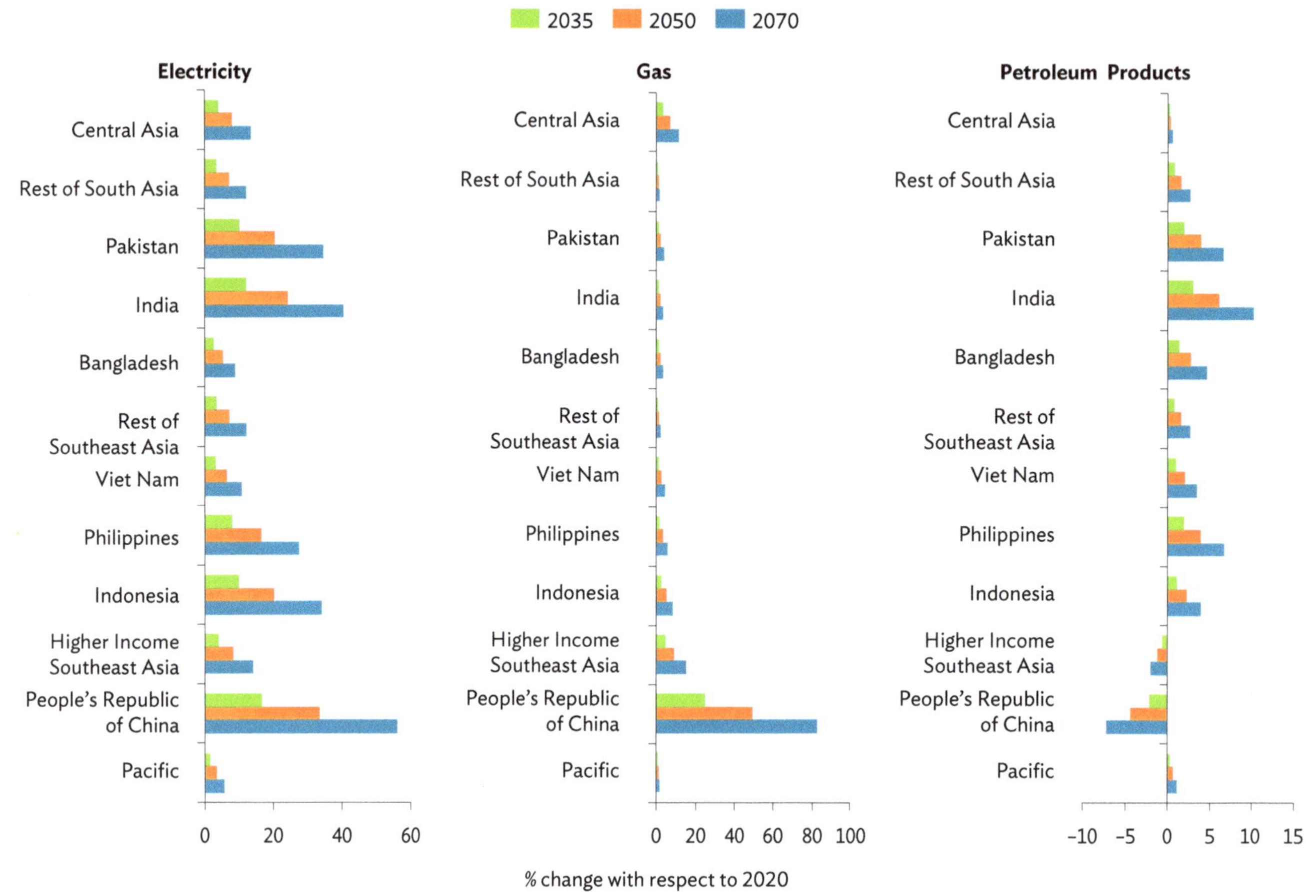

Notes: The scenario depicted is SSP5-8.5. Central Asia comprises Armenia, Azerbaijan, Georgia, Kazakhstan, the Kyrgyz Republic, Mongolia, Tajikistan, Turkmenistan, and Uzbekistan. Higher Income Southeast Asia comprises Brunei Darussalam, Malaysia, Singapore, and Thailand. The Pacific comprises the Cook Islands, the Federated States of Micronesia, Fiji, Kiribati, and the Marshall Islands. Rest of South Asia comprises Afghanistan, Bhutan, Maldives, Nepal, and Sri Lanka. Rest of Southeast Asia comprises Cambodia, the Lao People's Democratic Republic, Nauru, Niue, Palau, Papua New Guinea, Samoa, Solomon Islands, Tonga, Tuvalu, and Vanuatu.
Source: Campagnolo, L., G. Mansi, F. Bosello, and D. A. Raitzer. 2024. Quantifying the Economic Costs of Climate Change Inaction for the Asia-Pacific Region. Background paper for the *Asia–Pacific Climate Report 2024: Catalyzing Finance and Policy Solutions*. Asian Development Bank (results drawn from the Co-designing the Assessment of Climate Change Costs Project).

Where cooling is not feasible or economical, there will be adverse effects on labor supply and productivity. Essentially, there is an optimal perceived temperature for labor to be conducted. Beyond this humidity-adjusted temperature, labor productivity (output from those engaged in labor) declines, as does labor supply (willingness to engage in labor). In 2024, more than 70% of the global workforce was exposed to excessive heat and this proportion was higher in Asia and the Pacific (ILO 2024). With longer and more frequent heat waves, much of Asia and the Pacific will face contractions in both labor productivity and supply, especially in sectors where personal cooling is more difficult, such as agriculture and construction. Losses to labor have been calculated via functions of laboratory performance under different humidity-adjusted temperatures as well as via empirical analysis of micro survey data.

Empirical studies show how hotter temperatures reduce labor productivity. Dasgupta et al. (2021), using empirical estimates from microdata, find that effective labor would decline by 18 percentage points in Asia under 3°C of warming. Drawing on coefficients estimated by the study for low- and high-exposure sectors, Figure 2.13 highlights the differentiated impact on labor depending on high and low sectoral exposure. In the case of sectors with low exposure, the shocks in 2070 range from –16.2% (Bangladesh) and –1.3% (Central Asia) with respect to a 2020 baseline. The shock is more than doubled in the sectors with high exposure, reaching –31.1% in Bangladesh with respect to 2020. In addition to losses of effective labor, increased heat stress will lead to more workplace accidents and injuries. Fatima et al. (2021) estimate that each degree of temperature over 21°C leads to a 1% increase in risk of accidents.

Figure 2.13: Climate Change Impacts on Labor Productivity Under a High End Emissions Scenario

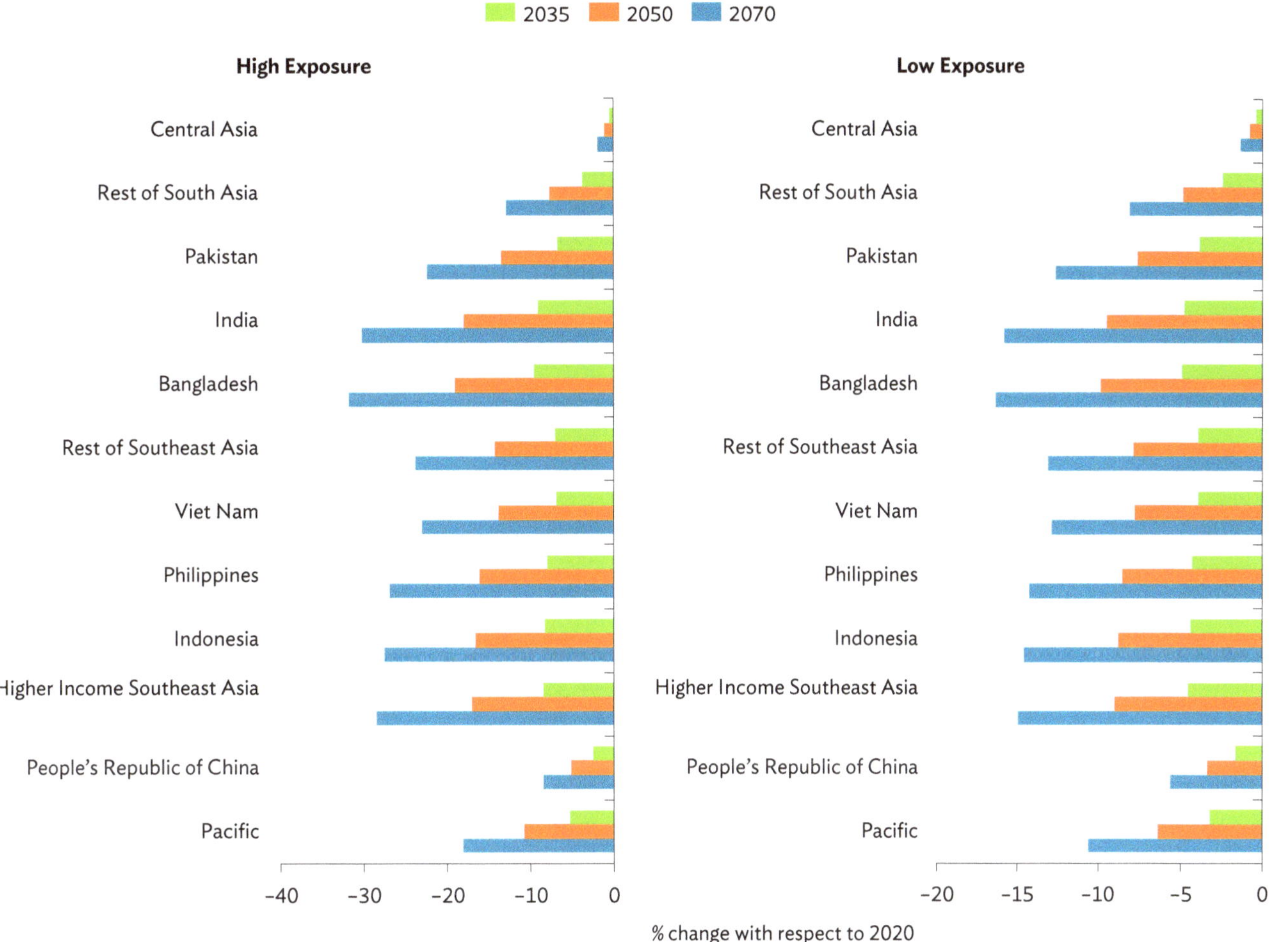

Notes: The scenario depicted is SSP5-8.5. Central Asia comprises Armenia, Azerbaijan, Georgia, Kazakhstan, the Kyrgyz Republic, Mongolia, Tajikistan, Turkmenistan, and Uzbekistan. Higher Income Southeast Asia comprises Brunei Darussalam, Malaysia, Singapore, and Thailand. The Pacific comprises the Cook Islands, the Federated States of Micronesia, Fiji, Kiribati, the Marshall Islands, Nauru, Niue, Palau, Papua New Guinea, Samoa, Solomon Islands, Tonga, Tuvalu, and Vanuatu. Rest of South Asia comprises Afghanistan, Bhutan, Maldives, Nepal, and Sri Lanka. Rest of Southeast Asia comprises Cambodia, the Lao People's Democratic Republic, Myanmar, and Timor-Leste.
Source: Campagnolo, L., G. Mansi, F. Bosello, and D. A. Raitzer. 2024. Quantifying the Economic Costs of Climate Change Inaction for the Asia-Pacific Region. Background paper for the *Asia–Pacific Climate Report 2024: Catalyzing Finance and Policy Solutions*. Asian Development Bank (results drawn from the Co-designing the Assessment of Climate Change Costs Project).

Climate change will adversely affect health in multiple ways. Accidents and injuries will increase due to increased flooding, landslides, and storms, while higher accident rates will occur under hotter, longer heat waves. Hotter peak temperatures increase cardiovascular risks and stress, particularly for those undertaking labor-intensive activities or for older people. Using empirical relationships between mortality and peak temperatures in conjunction with climate projections, Carleton et al. (2022) project that a high end emissions scenario could triple mortality due to heat waves, with heat waves causing as much mortality as all infectious diseases by 2100. The increases are more pronounced in lower-income and hotter countries, including much of Asia and the Pacific (Carleton et al. 2022). Vector-borne diseases will spread further and more quickly as mosquitoes and other vectors have larger ranges under climate change. For example, the population at risk of dengue is expected to rise by 2 billion globally under a high end emissions scenario, with faster transmission over a longer portion of each year in much of the Asia and Pacific region (Wang et al. 2024).

Climate change impacts will be felt most intensely by poor people. Climate change will have its strongest impacts on those with the most exposure and the lowest capacity to cope and adapt. A large share of poor people in the Asia and Pacific region are dependent on agriculture, fishing, or forestry, which are directly exposed to climate. Poor people often live in areas exposed to climatic hazards such as flooding, storm surges, or landslides; work in sectors exposed to heat stress; and/or have limited ability to afford improved cooling. Those with low incomes face greater risks from criminality and violence, and they have lower educational attainment—and both crime levels and disruptions to education will be exacerbated by longer hot periods. With poorer health, greater exposure to heat-related health risks, and less ability to afford health care, poor people will bear a disproportionate burden of climate-related morbidity and mortality. By some estimates, extreme poverty could be increased by 64%–117% by 2030 under a high end emissions climate scenario, relative to no climate change in Asia and the Pacific (Jafino et al. 2020).

2.4 Economic Impacts of Climate Change

Collectively, climate effects will have major economic impacts, but quantification is challenging. Many of the most important likely impacts of climate change are the most difficult to model, as they arise from changes to patterns of extreme events as well as complex interactions. GCMs agree on the big picture that extreme events will increase, but they often disagree on the details of exactly where and how. Traditionally, neoclassical integrated assessment models have represented the impacts of climate change as power functions of damages, but the science of the damages often has been out of date and focused on effects of a few changes, such as temperature and rainfall at the mean, rather than at the extremes (Calel and Stainforth 2017). As a result, economic damages due to climate change have often been earlier estimated as not more than a few years of gross domestic product (GDP) growth being lost by 2100.

Empirical studies have shown the potential for large economic impacts, but they have important limitations. Various econometric studies estimate relations between weather variations and economic activity, with the estimated coefficients applied to future climate scenarios to project future losses (Chang, Mi, and Wei 2023). These methods generate large economic losses (up to 40% of global GDP by 2100) if economic growth is taken as the dependent variable in the regressions, but more modest losses are anticipated if levels of economic activity are used. However, the larger loss estimates using growth rates also are sensitive to specification (Newell, Prest, and Sexton 2021). All econometric approaches also proxy weather for climate in a way that may overestimate losses, as there is less ability to foresee and adapt to weather variations than changes in climatic averages.

Innovative economy-wide modeling offers new insights on the consequences of climate change. This chapter has applied the latest science on shocks by sector in a global economy-wide model depicting impacts of a high end emissions scenario on the Asia and Pacific region. This approach avoids the limitations of econometric studies and builds from understanding of sector shocks to analysis of how entire economies are affected. The Intertemporal Computable Equilibrium System (ICES) is a recursive-dynamic multiregional Computable General Equilibrium model developed to assess economy-wide impacts of climate change on the economic system and to study climate change mitigation and adaptation policies. The general equilibrium structure of the ICES model allows for the analysis of market flows within each national economy and international flows with the rest of the world. This goes beyond the "simple" quantification of direct costs of a shock to offer an economic evaluation of secondary and higher-order effects within specific scenarios of climate change (Eboli, Parrado, and Roson 2010).

Shocks depicted in Figures 2.4–2.13 have been applied in the ICES model. This includes shocks to the agriculture, forestry, and fisheries sectors, increases to energy demand, reductions in labor productivity, and shocks created by increased flooding and sea-level rise (Figure 2.14). These are shocks with some forms of static infrastructure adaptation incorporated (such as coastal improvements) whereas other forms of market-based adaptation (such as crop substitution or movement of labor between sectors) are subsequently resolved in the economic model. In the case of agriculture, shocks on sector-specific land productivity are applied to modeled crops and extrapolated to other similar crops and livestock. For the forestry and fishing sectors, impacts are directly imputed as shocks on the sector-specific productivity of natural resources. Energy demand shocks are implemented as consumption shifts, in which households are forced to consume a different quantity of gas, electricity, and petroleum products, altering the demand and price of these commodities and also changing the consumption of all other goods due to the budget constraint. Climate-change-related temperature shocks on labor productivity are differentiated depending on the exposure of sectors, with agriculture, forestry, and fishing considered high exposure, industry and manufacturing considered low exposure, and the services sector not shocked. Impacts due to floods and sea-level rise are composite because they imply the simultaneous occurrence of a shock on labor productivity and a shock on capital stock. This shock on labor productivity is based on a temporary reduction during flood events, whereas the shock on capital stock is computed using the sectoral expected annual damage as a share of capital. In addition, sea-level rise shocks on land availability are included.

Figure 2.14: Framework to Quantify the Economic Effects of Climate Change Impacts

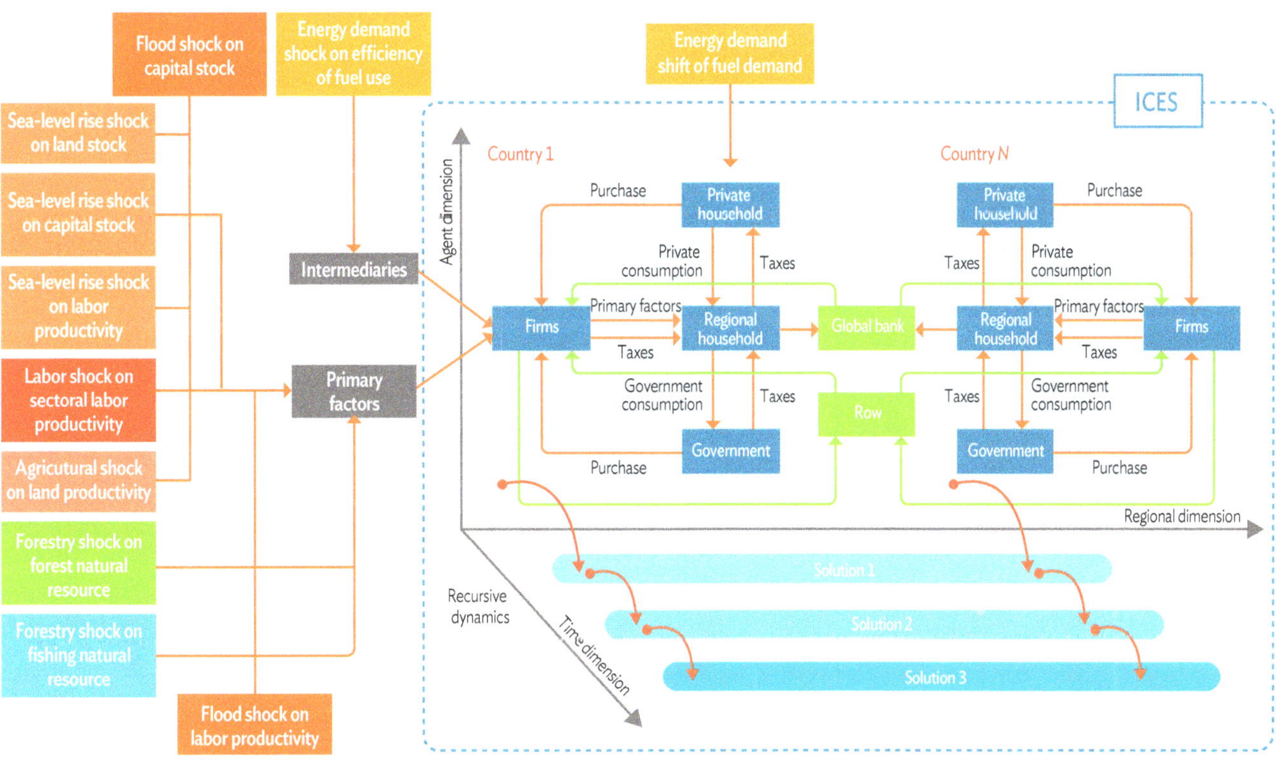

ICES = Intertemporal Computable Equilibrium System.
Source: Campagnolo, L., G. Mansi, F. Bosello, and D. A. Raitzer. 2024. Quantifying the Economic Costs of Climate Change Inaction for the Asia-Pacific Region. Background paper for the *Asia–Pacific Climate Report 2024: Catalyzing Finance and Policy Solutions*. Asian Development Bank.

The approach applied reflects a more appropriate representation of climate change shocks in economic modeling than typified earlier literature. Rather than treat climate change as creating recurrent shocks to output, climate change is also treated as creating persistent shocks to productivity and capital, which accumulate over time by changing the trajectory of growth. Multiple channels of impact are also considered, such as flood impacts on capital and labor productivity. The approach also includes impacts that occur via value chains, trade and investment, and competitiveness, as the model captures climate impacts in all regions of the globe, even if the focus here is on results for the Asia and Pacific region.

By 2070, climate change under a high end emissions scenario could cause a total loss of 16.9% of GDP across the Asia and Pacific region. Most of the region would face more than 20% loss. Aggregate results are presented in Figure 2.15 by country, while the Annex provides detail on the evolution of losses by sector in each country. Among the assessed countries and subregions, these losses are concentrated in Bangladesh (–30.5%), Viet Nam (–30.2%), Indonesia (–26.8%), India (–24.7%), "the rest of Southeast Asia" (–23.4%), higher-income Southeast Asia (–22.0%), Pakistan (–21.1%), the Pacific (–18.6%), and the Philippines (–18.1%). Losses increase at a higher rate over time, as the difference between 2050 and 2070 losses is greater than between 2030 and 2050. These losses are far above prior model-based losses and are consistent with the upper bound of econometric estimates. They also confirm that climate policy responses, including adaptation and mitigation, will be essential to the future welfare of the Asia and Pacific region.

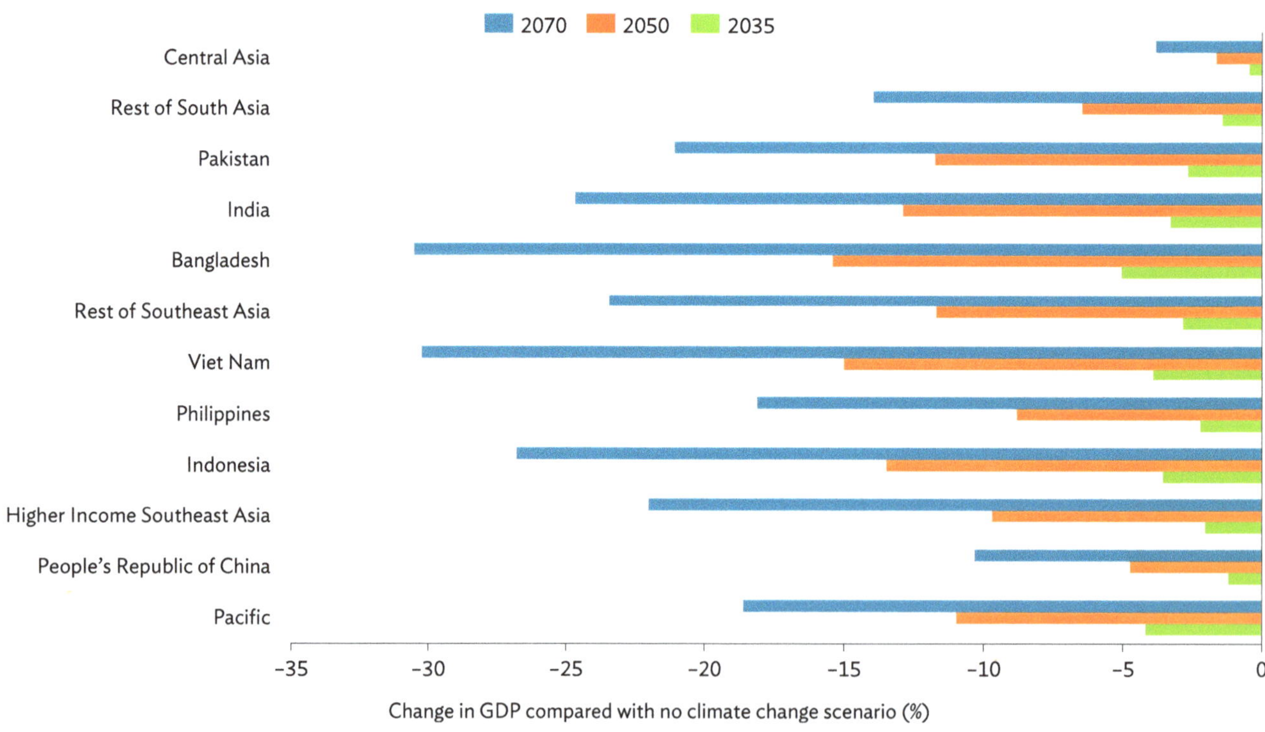

Figure 2.15: Loss of Gross Domestic Product due to Climate Change Under a High End Emissions Scenario

GDP = gross domestic product.
Notes: The scenario depicted is SSP5-8.5. Central Asia comprises Armenia, Azerbaijan, Georgia, Kazakhstan, the Kyrgyz Republic, Mongolia, Tajikistan, Turkmenistan, and Uzbekistan. Higher Income Southeast Asia comprises Brunei Darussalam, Malaysia, Singapore, and Thailand. The Pacific comprises the Cook Islands, the Federated States of Micronesia, Fiji, Kiribati, the Marshall Islands, Nauru, Niue, Palau, Papua New Guinea, Samoa, Solomon Islands, Tonga, Tuvalu, and Vanuatu. Rest of South Asia comprises Afghanistan, Bhutan, Maldives, Nepal, and Sri Lanka. Rest of Southeast Asia comprises Cambodia, the Lao People's Democratic Republic, Myanmar, and Timor-Leste.
Source: Campagnolo, L., G. Mansi, F. Bosello, and D. A. Raitzer. 2024. Quantifying the Economic Costs of Climate Change Inaction for the Asia-Pacific Region. Background paper for the *Asia–Pacific Climate Report 2024: Catalyzing Finance and Policy Solutions*. Asian Development Bank.

Losses could be even higher. Losses of 16.9% of regional GDP by 2070 are calculated under median sectoral shock values, but sector shocks have uncertainties, even under a single emissions scenario (Figure 2.16). If the 90th percentile of these shocks are considered, losses equate to 26.7% of regional GDP, a value that is 58.0% higher than under median shock values, with Bangladesh experiencing a 68.6% loss and a number of other locations facing losses of around 40%. Under the 10th percentile of shocks, the loss is 11.4% of regional GDP, which is still substantial. The estimates can also be considered conservative in that the modelling only reflected a subset of actual climate change impacts. For example, effects on human health beyond labor productivity are not included, nor are effects on ecosystem services, such as pollination, regulation of pests and diseases, or human conflict. In general, these omissions mean that a more comprehensive accounting would find even higher losses.

Sea-level rise is the dominant driver of economic losses due to climate change. Considering the impacts of sea-level rise and storm surges under a high end emissions climate scenario, the loss of GDP in 2070 for the Asia and Pacific region is 6.3% with respect to the "no climate change" scenario. Indonesia, Bangladesh, Viet Nam, the Pacific, and higher income Southeast Asia have the largest losses among developing countries and subregions at –16.1%,–15.4%, –15.0%, –13.8%, and –13.3% of GDP, respectively (Figure 2.16).

Figure 2.16: Sensitivity of Modeled Losses in 2070 due to Climate Change Under a High End Emissions Scenario

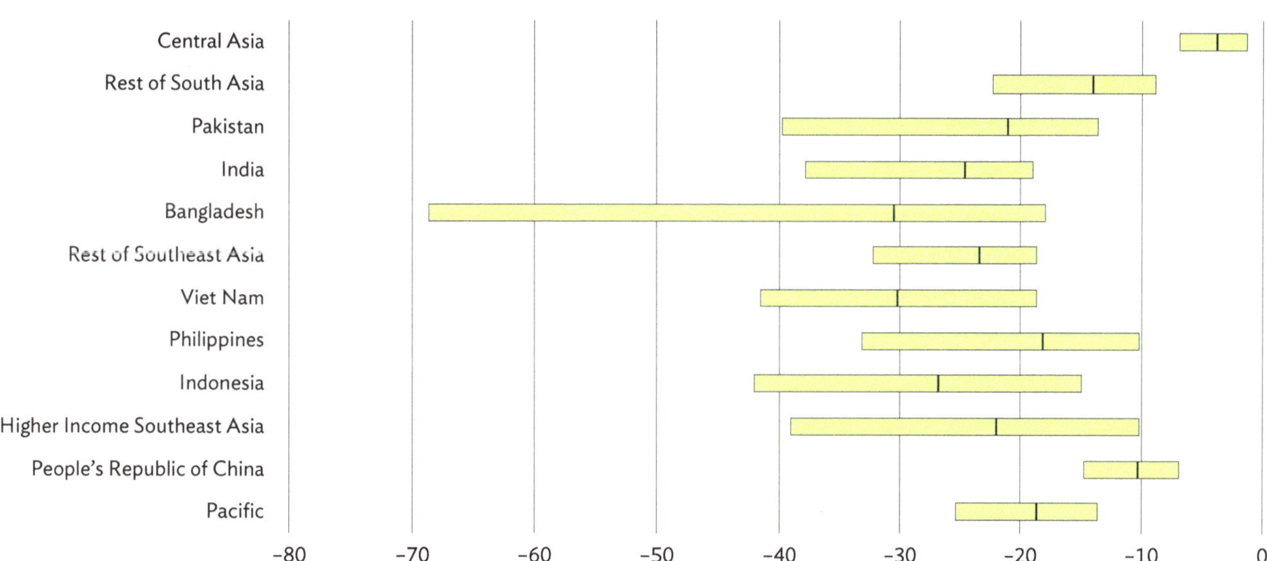

GDP = gross domestic product.
Notes: Bars depict the losses under the 10%–90% range of shocks, with the central line showing losses under median shocks. The scenario depicted is SSP5-8.5. Central Asia comprises Armenia, Azerbaijan, Georgia, Kazakhstan, the Kyrgyz Republic, Mongolia, Tajikistan, Turkmenistan, and Uzbekistan. Higher Income Southeast Asia comprises Brunei Darussalam, Malaysia, Singapore, and Thailand. The Pacific comprises the Cook Islands, the Federated States of Micronesia, Fiji, Kiribati, the Marshall Islands, Nauru, Niue, Palau, Papua New Guinea, Samoa, Solomon Islands, Tonga, Tuvalu, and Vanuatu. Rest of South Asia comprises Afghanistan, Bhutan, Maldives, Nepal, and Sri Lanka. Rest of Southeast Asia comprises Cambodia, the Lao People's Democratic Republic, Myanmar, and Timor-Leste.
Source: Campagnolo, L., G. Mansi, F. Bosello, and D. A. Raitzer. 2024. Quantifying the Economic Costs of Climate Change Inaction for the Asia-Pacific Region. Background paper for the *Asia–Pacific Climate Report 2024: Catalyzing Finance and Policy Solutions*. Asian Development Bank.

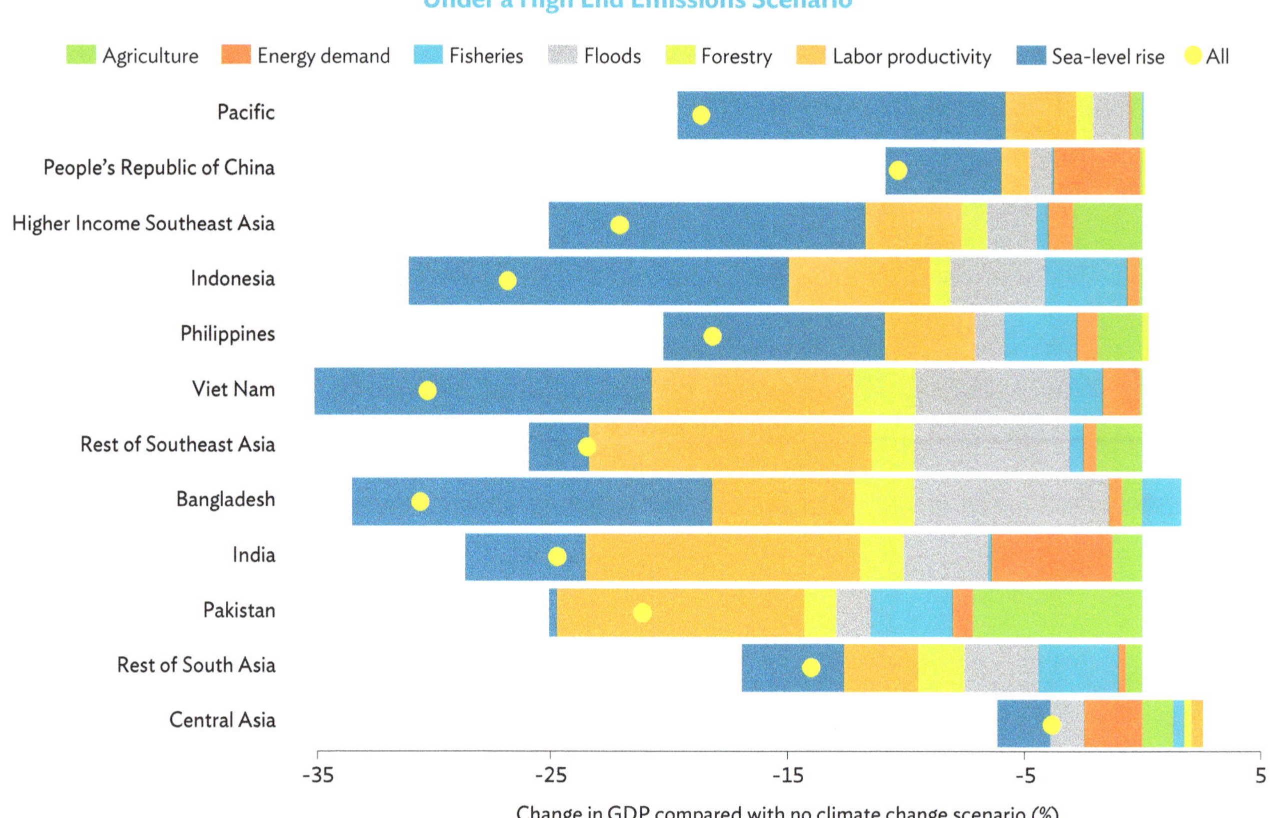

Figure 2.17: Composition of Losses in Gross Domestic Product in 2070 due to Climate Change Under a High End Emissions Scenario

GDP = gross domestic product.
Notes: The scenario depicted is SSP5-8.5. Central Asia comprises Armenia, Azerbaijan, Georgia, Kazakhstan, the Kyrgyz Republic, Mongolia, Tajikistan, Turkmenistan, and Uzbekistan. Higher Income Southeast Asia comprises Brunei Darussalam, Malaysia, Singapore, and Thailand. The Pacific comprises the Cook Islands, the Federated States of Micronesia, Fiji, Kiribati, the Marshall Islands, Nauru, Niue, Palau, Papua New Guinea, Samoa, Solomon Islands, Tonga, Tuvalu, and Vanuatu. Rest of South Asia comprises Afghanistan, Bhutan, Maldives, Nepal, and Sri Lanka. Rest of Southeast Asia comprises Cambodia, the Lao People's Democratic Republic, Myanmar, and Timor-Leste.
Source: Campagnolo, L., G. Mansi, F. Bosello, and D. A. Raitzer. 2024. Quantifying the Economic Costs of Climate Change Inaction for the Asia-Pacific Region. Background paper for the *Asia–Pacific Climate Report 2024: Catalyzing Finance and Policy Solutions*. Asian Development Bank.

For many countries, effects on labor productivity from heat waves are large and rank second or even dominate losses. The GDP loss in 2070 from reduced labor productivity is found to be 4.9% for the region. The most impacted locations are tropical and subtropical. These include "the rest of Southeast Asia" (–11.9%), India (–11.6%), Pakistan (–10.4%), and Viet Nam (–8.5%) as shown in Figure 2.17.

Increased energy demand becomes a substantial economic cost as economies struggle to cope with heat waves. Higher temperatures increase demand for cooling sufficiently to lead to a 3.3% GDP reduction for the Asia and Pacific region by 2070. Losses are concentrated in climates with hotter seasonal temperatures and where energy usage for cooling is highest. These include India (–5.1%), the PRC (–3.7%), and Central Asia (–2.5%), as shown in Figure 2.17.

Increased river-based flooding causes large economic impacts in selected countries and subregions. Due to increased river-based flooding under a high end emissions climate scenario, the GDP loss in 2070 for Asia and the Pacific is 2.2% of GDP. Countries with mega deltas experience the largest losses, with Bangladesh, "the rest of Southeast Asia," and Viet Nam facing changes of –8.2%, –6.6%, and –6.5% of GDP, respectively (Figure 2.17). Indonesia and India lose about 4% of GDP.

Losses are substantial in the natural resource sector, encompassing agriculture, forestry, and fisheries.
The economic loss from shocks to these climate-dependent sectors is 2.1% of GDP across the region by 2070.
The combined losses from these sectors are largest in Pakistan (–12.0%), "the rest of South Asia" (–6.0%),
the Philippines (–4.7%), higher income Southeast Asia (–4.5%), Indonesia (–4.5%), "the rest of Southeast Asia"
(–4.3%), and Viet Nam (-4.1%), as shown in Figure 2.17.

The composition of losses changes over time. Whereas sea-level rise is the largest source of economic
losses in 2070, in the 2030s, most economic losses are driven by effects on labor productivity and energy
demand (Figure 2.18). Losses related to sea-level rise and storm surges accelerate over time, because of a
combined effect of sea-level rise itself accelerating, storm surges exacerbating the acceleration effect, and the
cumulative effect on economic growth of costs to rebuild damaged capital. The modeling also finds smaller
combined effects of shocks than the shocks considered individually, as some shocks affect the same sectors
and there is compensation within the economy.

Figure 2.18: Sector Composition of Modeled Losses due to Climate Change Under a High End Emissions Scenario

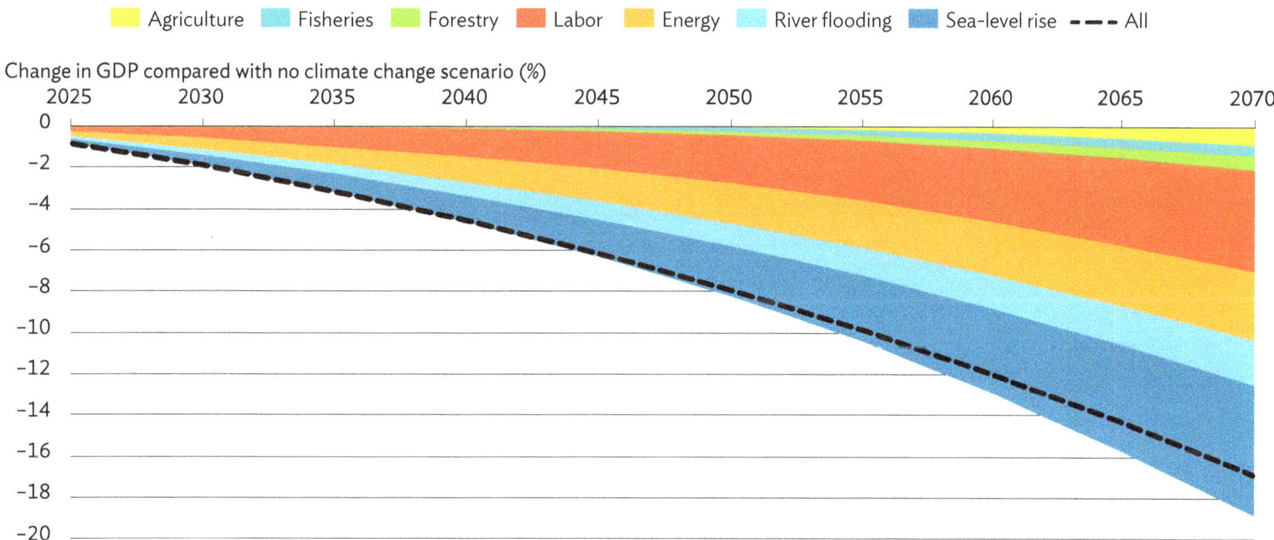

GDP = gross domestic product.
Notes: The scenario depicted is SSP5-8.5. The figure includes Afghanistan, Armenia, Azerbaijan, Bangladesh, Bhutan, Brunei Darussalam, Cambodia,
the People's Republic of China, the Cook Islands, Fiji, Georgia, India, Indonesia, Kazakhstan, Kiribati, the the Kyrgyz Republic, the Lao People's
Democratic Republic, Malaysia, Maldives, the Marshall Islands, the Federated States of Micronesia, Mongolia, Myanmar, Nauru, Nepal, Niue, Pakistan,
Palau, Papua New Guinea, the Philippines, Samoa, Singapore, Sri Lanka, Solomon Islands, Tajikistan, Thailand, Timor-Leste, Tonga, Turkmenistan,
Tuvalu, Uzbekistan, Vanuatu, and Viet Nam.
Source: Campagnolo, L., G. Mansi, F. Bosello, and D. A. Raitzer. 2024. Quantifying the Economic Costs of Climate Change Inaction for the
Asia-Pacific Region. Background paper for the *Asia–Pacific Climate Report 2024: Catalyzing Finance and Policy Solutions.* Asian Development Bank

Damage functions derived from the modeling suggest that 41.0% of GDP could be lost by 2100.
Although economic losses are only directly modeled out to 2070, the relationships between climate
variables and losses up to 2070 can be used to extrapolate losses to 2100. This involves fitting country- or
subregion-specific damage functions of GDP loss to global mean temperature increase for loss channels via
labor productivity, energy demand, river-based flooding, and fisheries and forestry shocks. For agriculture,
GDP loss is fitted to the global mean SPEI-6 change, while GDP loss due to sea-level rise is fitted to mean
global sea-level rise. All shocks are rescaled to account for the fact that simultaneous runs of shocks lead to
smaller losses than the sum of losses from shocks modeled individually. These damage functions are applied
to the mean of 25 GCM projections from the Coupled Model Intercomparison Project 6 to extrapolate losses
to 2100 for temperature and SPEI and Garner et al. (2021) sea-level rise projections for the IPCC Sixth
Assessment Report (Figure 2.19). The results for a high end emissions climate scenario (SSP5-8.5) show

especially large vulnerability for Bangladesh, with a potential 77.9% GDP reduction, Viet Nam with 67.9% reduction, and Indonesia with 61.1% GDP reduction. The result of 41.0% GDP loss for the Asia and Pacific region is higher than previous estimates of 24% of GDP (ADB 2023) and less than 15% of GDP for subregions of Asia (Raitzer et al. 2015; Ahmed and Suphachalasai 2014; Westphal, Hughes, and Brömmelhörster 2013).

Losses remain substantial even under much more optimistic mitigation assumptions. The same damage functions and data sources that extrapolate losses to 2100 under a high end emissions scenario are applied to other scenarios, including RCP4.5,[5] which leads to warming of about 2.7°C, and RCP2.6 as a scenario that keeps global warming well below 2°C. Under RCP4.5, regional losses would be 20.0% of GDP in 2100, with Bangladesh experiencing 41.1% GDP loss, Viet Nam losing 33.8%, and a number of other countries facing 20%–30% losses (Figure 2.19). Under RCP2.6, which would require immediate emissions reductions that reach global net zero by 2070, regional losses would be 11.2% of GDP, with Bangladesh still facing a 24.5% GDP loss and several other countries and subregions facing losses in excess of 15% of GDP. While this is a major reduction from losses under RCP8.5, the result also illustrates that substantial climate risks remain to be addressed by adaptation measures, even if mitigation is pursued with high ambition and no climate tipping points are triggered.

Figure 2.19: Economic Losses in 2100 due to Climate Change Under Various Climate Scenarios

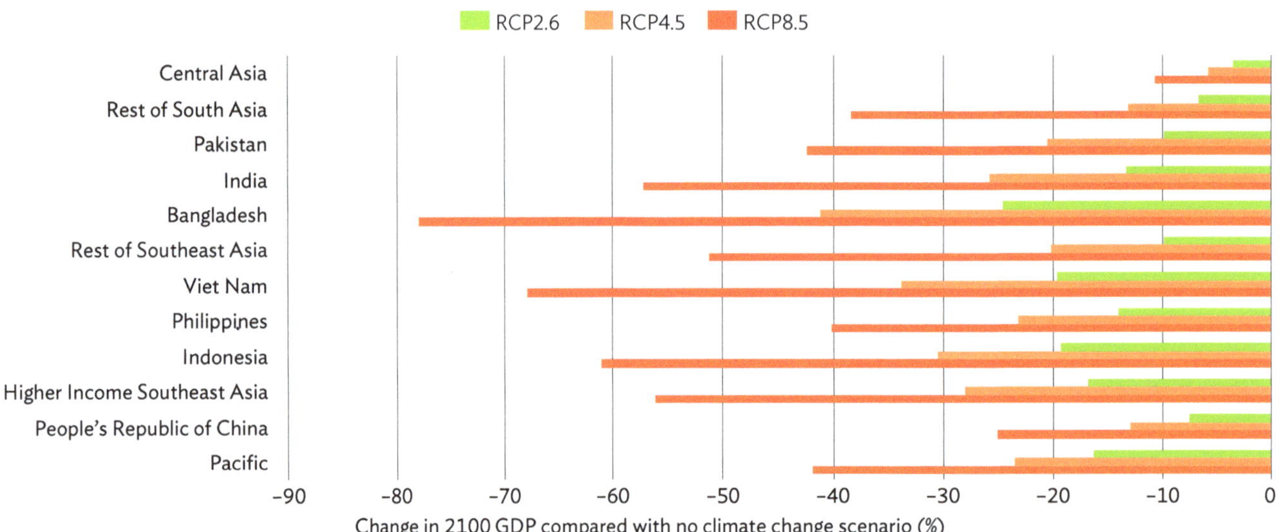

GDP = gross domestic product, RCP = Representative Concentration Pathway.
Notes: Central Asia comprises Armenia, Azerbaijan, Georgia, Kazakhstan, the Kyrgyz Republic, Mongolia, Tajikistan, Turkmenistan, and Uzbekistan. Higher Income Southeast Asia comprises Brunei Darussalam, Malaysia, Singapore, and Thailand. The Pacific comprises the Cook Islands, the Federated States of Micronesia, Fiji, Kiribati, the Marshall Islands, Nauru, Niue, Palau, Papua New Guinea, Samoa, Solomon Islands, Tonga, Tuvalu, and Vanuatu. Rest of South Asia comprises Afghanistan, Bhutan, Maldives, Nepal, and Sri Lanka. Rest of Southeast Asia comprises Cambodia, the Lao People's Democratic Republic, Myanmar, and Timor-Leste.
Source: Campagnolo, L., G. Mansi, F. Bosello, and D. A. Raitzer. 2024. Quantifying the Economic Costs of Climate Change Inaction for the Asia-Pacific Region. Background paper for the *Asia–Pacific Climate Report 2024: Catalyzing Finance and Policy Solutions*. Asian Development Bank.

Overall patterns reinforce that economic impacts will be regressive. With the largest economic impacts of climate change occurring in those areas of Asia and the Pacific with lower relative incomes—such as Bangladesh, India, Indonesia, Pakistan, the rest of Southeast Asia, Viet Nam, and the Pacific—via channels such as increased coastal inundation, lower labor productivity, and lower natural resource productivity, poor communities are likely to be most affected. Poor people are most likely to live in areas exposed to flooding and storm surges, to depend on natural resource sectors for livelihoods, and/or to work in sectors exposed to increasing heat. They also have the lowest capacity to adapt and cope with these shocks.

---

[5] The text here refers to RCPs, rather than the SSP designation (e.g., RCP8.5, rather than SSP5-8.5), because the socioeconomic assumptions are held constant in the comparison, so that designations such as SSP2-4.5 and SSP1-2.6 are not correct for those scenarios.

Annex: Details of Modeled Losses of Gross Domestic Product Under a High End Emissions Scenario

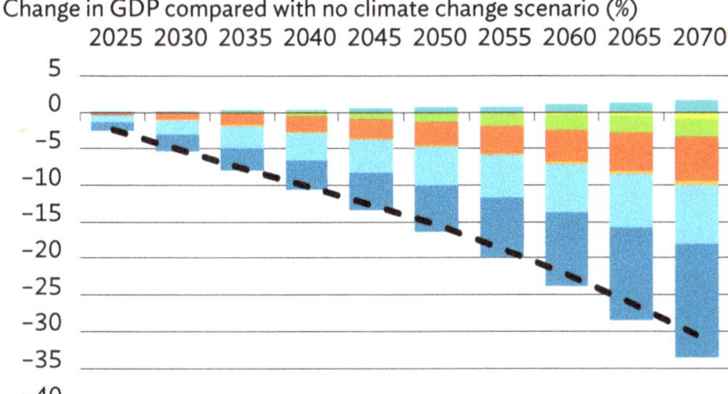

Bangladesh

Change in GDP compared with no climate change scenario (%)

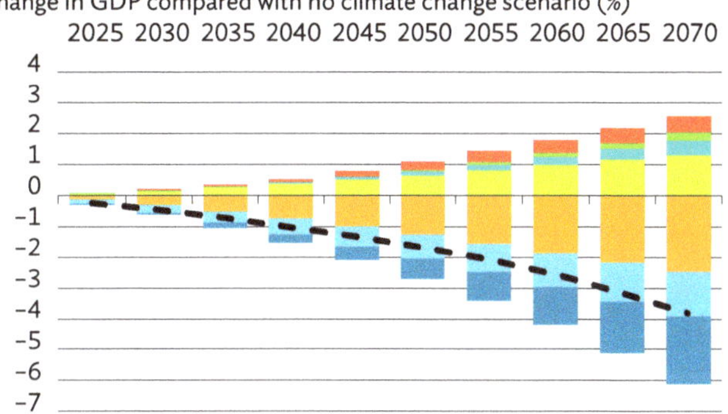

Central Asia

Change in GDP compared with no climate change scenario (%)

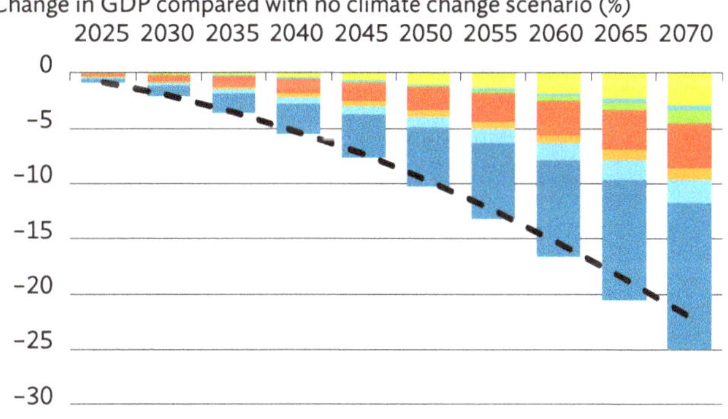

People's Republic of China

Change in GDP compared with no climate change scenario (%)

Higher Income Southeast Asia

Change in GDP compared with no climate change scenario (%)

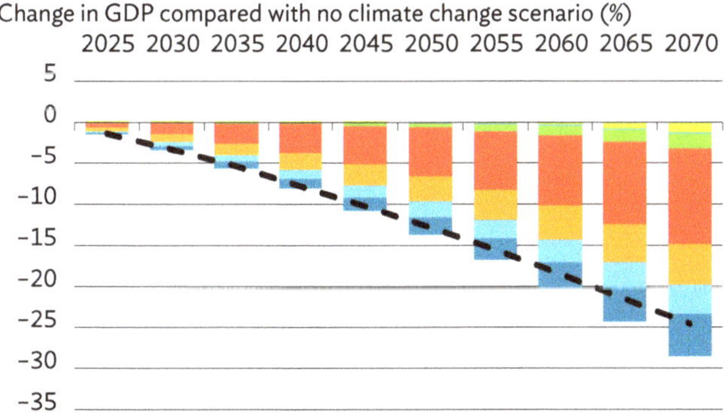

India

Change in GDP compared with no climate change scenario (%)

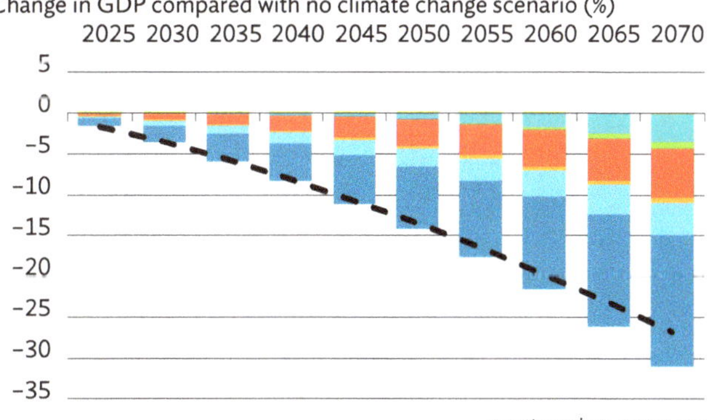

Indonesia

Change in GDP compared with no climate change scenario (%)

continued on next page

Annex *continued*

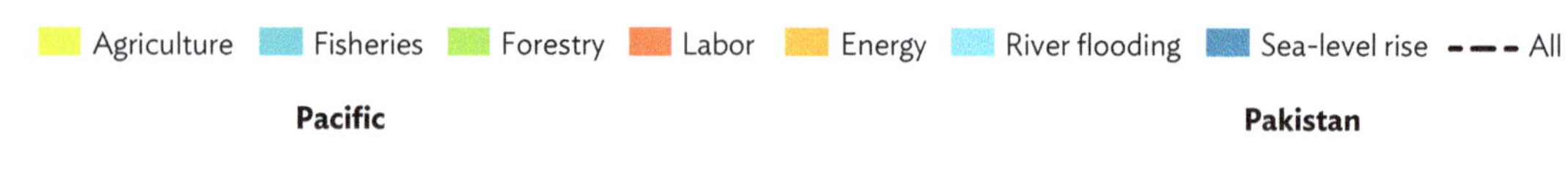

Pacific

Change in GDP compared with no climate change scenario (%)

Pakistan

Change in GDP compared with no climate change scenario (%)

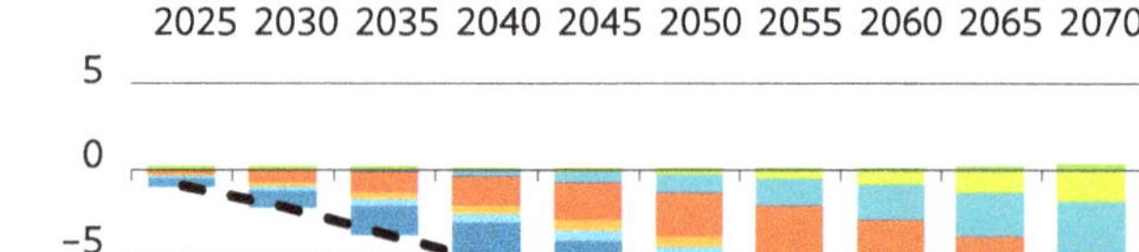

Philippines

Change in GDP compared with no climate change scenario (%)

Rest of South Asia

Change in GDP compared with no climate change scenario (%)

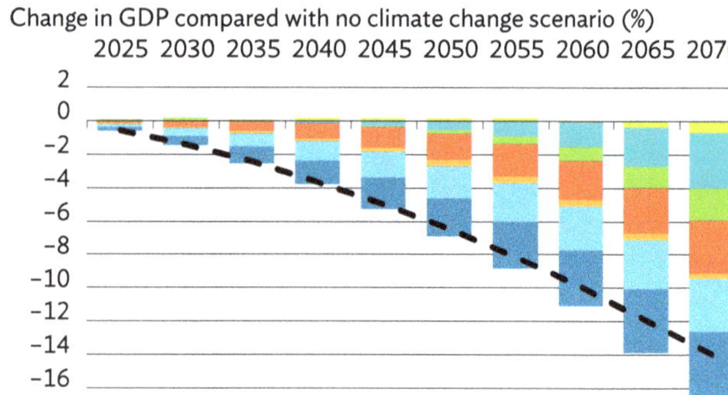

Rest of Southeast Asia

Change in GDP compared with no climate change scenario (%)

Viet Nam

Change in GDP compared with no climate change scenario (%)

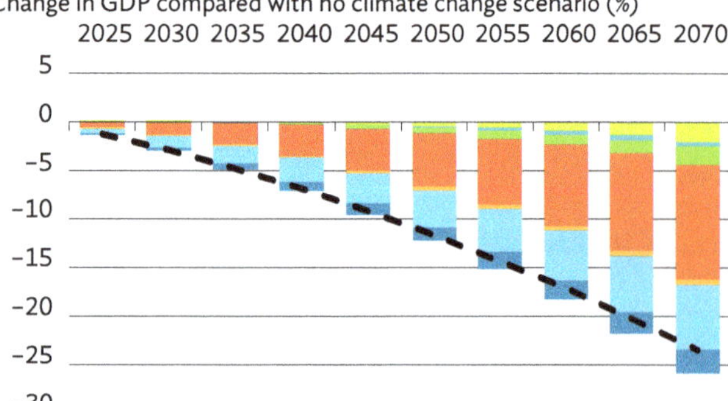

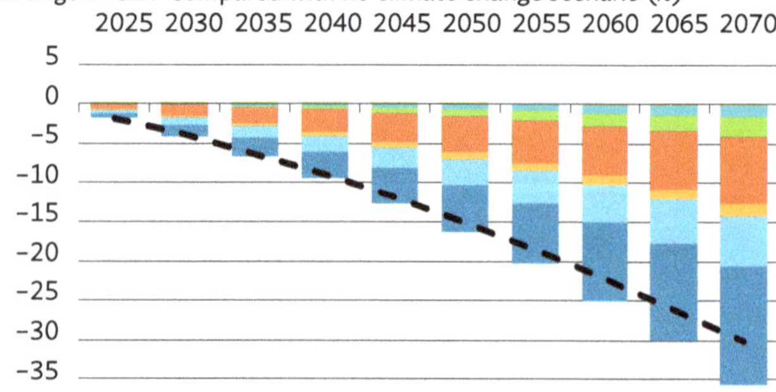

GDP = gross domestic product.
Notes: The scenario depicted is SSP5-8.5. Central Asia comprises Armenia, Azerbaijan, Georgia, Kazakhstan, the Kyrgyz Republic, Mongolia, Tajikistan, Turkmenistan, and Uzbekistan. Higher Income Southeast Asia comprises Brunei Darussalam, Malaysia, Singapore, and Thailand. The Pacific comprises the Cook Islands, the Federated States of Micronesia, Fiji, Kiribati, the Marshall Islands, Nauru, Niue, Palau, Papua New Guinea, Samoa, Solomon Islands, Tonga, Tuvalu, and Vanuatu. Rest of South Asia comprises Afghanistan, Bhutan, Maldives, Nepal, and Sri Lanka. Rest of Southeast Asia comprises Cambodia, the Lao People's Democratic Republic, Myanmar, and Timor-Leste.
Source: Campagnolo, L., G. Mansi, F. Bosello, and D. A. Raitzer. 2024. Quantifying the Economic Costs of Climate Change Inaction for the Asia-Pacific Region. Background paper for the *Asia–Pacific Climate Report 2024: Catalyzing Finance and Policy Solutions*. Asian Development Bank.

References

ADB (Asian Development Bank). 2024. *Asia in the Global Transition to Net Zero: Asian Development Outlook 2023 Thematic Report.*

Ahmed, M. and S. Suphachalasai. 2014. *Assessing the Costs of Climate Change and Adaptation in South Asia.* ADB.

Becker, M., M. Karpytchev, and A. Hu. 2023. Increased Exposure of Coastal Cities to Sea-Level Rise due to Internal Climate Variability. *Nature Climate Change.* 13. pp. 367–374.

Bjordal, J., T. Storelvmo, K. Alterskjær, and T. Carlsen. 2020. Equilibrium Climate Sensitivity Above 5 °C Plausible due to State-Dependent Cloud Feedback. *Nature Geoscience.* 13. pp. 718–721.

Burke, L., K. Reytar, M. Spalding, and A. Perry. 2012. *Reefs at Risk Revisited: Summary for Decision Makers.* World Resources Institute.

C3S (Copernicus Climate Change Service). 2024. Copernicus: February 2024 Was Globally the Warmest on Record – Global Sea Surface Temperatures at Record High. 5 March.

Calel, R. and D. Stainforth. 2017. On the Physics of Three Integrated Assessment Models. *American Meteorological Society.* 98 (6). pp. 1199–1216.

Carleton, T., A. Jina, M. Delgado, M. Greenstone, T. Houser, S. Hsiang, A. Hultgren, et al. 2022. Valuing the Global Mortality Consequences of Climate Change Accounting for Adaptation Costs and Benefits. *Climate Impact Lab.* 1 November.

Chang, J., Z. Mi, and Y. Wei. 2023. Temperature and GDP: A Review of Climate Econometrics Analysis. *Structural Change and Economic Dynamics.* 66. pp. 383–392.

Chen, J., C. Y. Tam, K. Cheung, Z. Wang, H. Murakami, N. C. Lau, S. T. Garner, et al. 2021. Changing Impacts of Tropical Cyclones on East and Southeast Asian Inland Regions in the Past and a Globally Warmed Future Climate. *Frontiers in Earth Science.* Interdisciplinary Climate Studies 9.

Chen, W. and W. Huan. 2019. Gaps Between Pre-2020 Climate Policies with NDC Goals and Long-Term Mitigation Targets: Analyses on Major Regions. *Energy Procedia.* 158. pp. 3664–3669.

CICA (Copernicus Interactive Climate Atlas). Copernicus Interactive Climate Atlas (accessed 5 May 2024).

Colucci, R. 2022. A Powerful Heatwave is Affecting a Large Part of Antarctica with Unprecedented Warm Temperatures More than 40 Degrees Above Average. Severe Weather Europe. 20 March.

Dasgupta, S., N. van Maanen, S. N. Gosling, F. Piontek, C. Otto, and C. F. Schleussner. 2021. Effects of Climate Change on Combined Labour Productivity and Supply: An Empirical, Multi-Model Study. *Elsevier.* 5 (7). pp. 455–465.

De Cian, E. and I. Sue Wing. 2019. Global Energy Consumption in a Warming Climate. *Environment and Resource Economics.* 72. pp. 365–410.

Duan, R., G. Huang, X. Zhou, Y. Li, and C. Tian. 2021. Ensemble Drought Exposure Projection for Multifactorial Interactive Effects of Climate Change and Population Dynamics: Application to the Pearl River Basin. *Earth's Future.* 9 (8).

Eboli, F., R. Parrado, and R. Roson. 2010. Climate Change Feedback on Economic Growth: Explorations with Dynamic Equilibrium Model. *Environmental and Development Economics.* 15 (5). pp. 515–533.

Fatima, S. H., P. Rothmore, L. C. Giles, B. M. Varghese, and P. Bi. 2021. Extreme Heat and Occupational Injuries in Different Climate Zones: A Systematic Review and Meta-Analysis of Epidemiological Evidence. *Environment International.* 148. 106384.

Garner, G. G., T. Hermans, R. E. Kopp, A. B. A. Slangen, T. L. Edwards, A. Levermann, S. Nowikci et al. 2021. Global Projection: Sea-Level Rise Under Different SSP Scenarios. Version 20210809 (accessed 8 January 2024).

Giardino, A. and M. Vousdoukas. 2024. Rising Seas: Building Resilience Against Coastal Flooding in Asia and the Pacific. *Development Asia.* 7 August.

Heinicke, S., K. Frieler, J. Jägermeyr, and M. Mengel. 2022. Global Gridded Crop Models Underestimate Yield Responses to Droughts and Heatwaves. *Environmental Research Letters.* 17 (4).

Hu, X., P. Chen, H. Huang, T. Sun, and D. Li. 2017. Contrasting Impacts of Heat Stress on Violent and Nonviolent Robbery in Beijing, China. *Natural Hazards.* 87. pp. 961–972.

IEA (International Energy Agency). 2019. The Future of Cooling in Southeast Asia.

IIASA (International Institute for Applied Systems Analysis). 2021. Global Forest Model (G4M).

ILO (International Labour Organization). 2024. Ensuring Safety and Health at Work in a Changing Climate.

IPCC (Intergovernmental Panel on Climate Change). 2022a. Climate Change 2022: Mitigation of Climate Change. Working Group III Contribution to the IPCC Sixth Assessment Report.

______. 2022b. Climate Change 2022: Impacts, Adaptation and Vulnerability.

ISIMIP (Inter-Sectoral Impact Model Intercomparison Project). ISIMIP Repository (accessed 1 March 2024).

Jafino, B. A., B. Walsh, J. Rozenberg, and S. Hallegatte. 2020. Revised Estimates of the Impact of Climate Change on Extreme Poverty by 2030. *Policy Research Working Paper.* No. 9417. World Bank.

Ke, P., P. Ciais, S. Sitch, W. Li, A. Bastos, Z. Liu, Y. Xu, et al. 2024. Low Latency Carbon Budget Analysis Reveals a Large Decline of the Land Carbon Sink in 2023. ArXiv. 10.48550/arXiv.2407.12447.

Kirschbaum, D., S. B. Kapnick, T. Stanley, and S. Pascale. 2020. Changes in Extreme Precipitation and Landslides Over High Mountain Asia. *Geophysical Research Letters.*

Kulp, S. and B. Strauss. 2019. New Elevation Data Triple Estimates of Global Vulnerability to Sea-Level Rise and Coastal Flooding. *Nature Communications.* 10 (4844).

NCEI (National Centers for Environmental Information). 2024. Annual 2023 Global Climate Report.

Newell, R. G., B. C. Prest, and S. E. Sexton. 2021. The GDP-Temperature Relationship: Implications for Climate Change Damages. *Journal of Environmental Economic Management.* 108. 102445.

Park, R., J. Goodman, M. Hurwitz, and J. Smith. 2020. Heat and Learning. *American Economic Journal: Economic Policy.* 12 (2). pp. 306–339.

Raitzer, D. A., F. Bosello, M. Tavoni, C. Orecchia, G. Marangoni, and J. N. G. Samson. 2015. *Southeast Asia and the Economics of Global Climate Stabilization.* ADB.

Raitzer, D. A. and J. Drouard. 2024. Empirically Estimated Impacts of Climate Change on Global Crop Production via Increasing Precipitation-Evapotranspiration Extremes. Background paper for the *Asia–Pacific Climate Report 2024: Catalyzing Finance and Policy Solutions.* ADB.

Ruane, A., M. Phillips, J. Jägermeyr, and C. Müller. 2024. Non-Linear Climate Change Impacts on Crop Yields May Mislead Stakeholders. Earth's Future.

Sanz-Barbero, B., C. Linares, C. Vives-Cases, J. L. González, J. J. López-Ossorio, and J. Díaz. 2018. Heat Wave and the Risk of Intimate Partner Violence. *Science of the Total Environment.* 644. pp. 413–419.

Schwalm, C. R., S. Glendon, and P. B. Duffy. 2020. RCP8.5 Tracks Cumulative CO_2 Emissions. *Proceedings of the National Academy of Sciences of the United States of America.* 117 (33). pp. 19656–19657.

SOCAT. 2024. Surface Ocean CO_2 Atlas.

Standardi, G., S. Dasgupta, R. Parrado, E. De Cian, and F. Bosello. 2023. Assessing Macro-Economic Effects of Climate Impacts on Energy Demand in EU Sub-National Regions. *Environmental and Resource Economics.* 86 (1). pp. 173–201.

van Vuuren, D. P., J. Lowe, E. Stehfest, L. Gohar, A. Hof, C. Hope, R. Warren, M. Meinshausen, and G. K. Plattner. 2009. How Well Do Integrated Assessment Models Simulate Climate Change? *Climatic Change.* 104 (2). pp. 255–285.

Venegas, R., J. Acevedo, and E. Treml. 2023. Three Decades of Ocean Warming Impacts on Marine Ecosystems: A Review and Perspective. *Deep Sea Research Part II: Topical Studies in Oceanography.* 212.

Vousdoukas, M. I., L. Mentaschi, E. Voukouvalas, M. Verlaan, S. Jevrejeva, L. P. Jackson, and L. Feyen. 2018. Global Probabilistic Projections of Extreme Sea Levels Show Intensification of Coastal Flood Hazard. *Nature Communications.* 9. 2360.

Wang, Y., C. Li, S. Zhao, Y. Wei, K. Li, X. Jiang, J. Ho, et al. 2024. Projection of Dengue Fever Transmissibility Under Climate Change in South and Southeast Asian Countries. *PLOS Neglected Tropical Diseases.* 29 April.

Ward, P., B. Jongman, J. C. J. H. Aerts, P. D. Bates, W. J. W. Botzen, A. Diaz Loaiza, S. Hallegatte, et al. 2017. A Global Framework for Future Costs and Benefits of River-flood Protection in Urban Areas. *Nature Climate Change.* 7. pp. 642–646.

Westphal, M., G. Hughes, and J. Brömmelhörster. 2013. *Economics of Climate Change in East Asia.* ADB.

Wijedasa, L., S. Sloan, S. Page, G. Clements, M. Lupascu, and T. Evans. 2018. Carbon Emissions from South-East Asian Peatlands Will Increase Despite Emission-Reduction Schemes. *Global Change Biology.* 24 (10). pp. 4598–4613.

Wilcox, P., M. Mudelsee, C. Spötl, and R. Lawrence Edwards. 2023. Solar Forcing of ENSO on Century Timescales. *Geophysical Research Letters.*

Winsemius, H. C., J. C. J. H. Aerts, L. P. H. van Beek, M. F. P. Bierkens, A. Bouwman, B. Jongman, J. C. J. Kwadijk, et al. 2017. Global Drivers of Future River Flood Risk. *Nature Climate Change.* 6. pp. 381–385.

WMO (World Meteorological Organization). 2024. WMO Confirms that 2023 Smashes Global Temperature Record. Press Release. 12 January.

You, X. 2023. Surge in Extreme Forest Fires Fuels Global Emissions. *Nature Climate Change.* 20 December.

Zhang, Z., B. Poulter, A. Feldman, Q. Ying, P. Ciais, S. Peng, and X. Li. 2023. Recent Intensification of Wetland Methane Feedback. *Nature Climate Change.* 13. pp. 430–433.

Chapter 3

Accelerating Climate Change Adaptation in Asia and the Pacific

3.1   Framing Adaptation Responses to Climate Change

A warmer and destabilized climate system necessitates urgent responses. Climate change is already irreversible, and patterns of warming will take decades to respond to mitigation actions. This means that climate adaptation is essential. Although mitigation will condition the amount of adaptation needed in the long term, both unavoidable climate impacts and the possibility that larger impacts will not be avoided must be tackled by adaptation measures in the near term. As illustrated in Chapter 2 of this report, the Asia and Pacific region faces large economic and social losses if climate change is not met with commensurate responses. A wide range of actions will be needed to avoid large losses from climate change that is locked in, as well as the likelihood that warming will continue to escalate. This chapter starts by providing a framing for how adaptation policy can be approached. It then reviews priority adaptation responses in vulnerable sectors and systems. Actions to date are compared against priority responses, with a particular focus on climate finance. Finally, recommendations are offered based on gaps between priority responses and actions to date. Recommendations to upscale adaptation efforts are framed around three sets of policy actions: (i) assessing and communicating information on climate risk to embed adaptation in many decisions; (ii) systematically planning, prioritizing, and appraising investments in adaptation; and (iii) securing and sustaining adaptation financing.

3.1.1   Adaptation Context

Effective adaptation has to respond to a mix of hazards, including slow-onset changes as well as changes in variability and extremes. Slow-onset events refer to risks and impacts that unfold gradually over time, such as the increase in mean temperature and changing precipitation or sea-level rise and salinization (van der Geest and van den Berg 2021). The impacts of these changes can be reduced with adaptation. Climate change will also lead to changes in the patterns of extreme weather, changing the frequency and/or intensity of events such as heavy precipitation and floods, droughts, storms (including cyclones and typhoons), and heat waves, with increased likelihood of unprecedented extremes (Seneviratne et al. 2021). Addressing these acute impacts of climate change involves a combination of disaster risk management and climate change adaptation.

Resilience to climate risk is a function of how exposure and vulnerability condition the effects of changing hazards. Hazards include slow-onset changes and extreme events driven by the biophysical processes of climate change. On the other hand, exposure and vulnerability are largely conditioned by socioeconomic processes and are also dynamic (IPCC 2014). This means that socioeconomic processes also drive climate risk, and effective adaptation therefore requires that these underlying drivers are addressed to reduce exposure and vulnerability.

Adaptation is a continual process, not a set of one-off decisions. The potential impacts of climate change—and the analysis of adaptation responses—are specific to risk, location, context, and time (UNEP 2023). These challenging factors are compounded by high levels of uncertainty associated with future climate change impacts and levels of associated risk and vulnerability. Uncertainty is evident in four main areas: (i) policy uncertainty about future emissions trajectories; (ii) model and scientific uncertainty about what emissions will mean for climate and weather patterns, especially in specific places and at particular times; (iii) biophysical impact uncertainty about how weather and climate shifts impact ecosystem-dependent goods and services; and (iv) socioeconomic uncertainty about how the future world will evolve and change exposure and vulnerability to biophysical impacts. This layering of evolving uncertainties means that adaptation is not a simple linear cost-optimization exercise and requires approaches to optimize decisions under uncertainty

This chapter was written by a team consisting of ADB staff Sabah Abdullah, Alessio Giardino, Neil Foster-McGregor, David A. Raitzer (lead author), and Arghya Sinha Roy, with consultant Paul Watkiss.

(Haasnoot et al. 2013). Adaptive management and adaptation pathways therefore should consider adaptation as a process, rather than a defined set of technical solutions. This adaptation "sequencing" builds in learning over time to allow for updated options as new information emerges.

Reducing the costs of climate change impacts through adaptation should have economic benefits.
While climate adaptation has an overall cost—defined as the combined costs of planning, preparing for, facilitating, and implementing adaptation measures—it seeks to reduce the economic and financial costs of climate change. Estimating the costs and benefits of adaptation requires a complex analysis of site-specific risks (hazard, vulnerability, and exposure), which may change over time. Estimates of adaptation costs can also vary according to climate objectives, projected warming levels and scenarios, definitions and boundaries, and the sectors and risks analyzed. However, appropriate adaptation measures should lead to net economic benefits. This is reinforced by economic assessments of individual solutions or adaptation projects, including reviews of the cost–benefit ratios of adaptation by the Organisation for Economic Co-operation and Development (OECD 2015), the Global Commission on Adaptation (GCA 2019, 2021), the Climate Change Committee (CCC 2021), and the World Bank (World Bank 2021). These studies report that many, though not all, adaptation options deliver positive cost–benefit ratios, typically above 2:1 and often as high as 10:1 (Giardino and Vousdoukas 2024). Because of its positive effects in reducing the fiscal impacts of climate change, adaptation can also have net benefits for public finances, even though it may require higher spending (Bachner, Bednar-Friedl, and Knittel 2019). Adaptation can prevent climate impacts that reduce government revenues, increase government expenditure and contingent liabilities, alter external performance, and even influence sovereign credit ratings and the cost of debt (UNEP 2018; Volz et al. 2020; Cevik and Jalles 2020).

Adaptation involves a trade-off between costs, benefits, and residual risk. Appropriate economic analysis of climate adaptation requires assessment of not only the costs and benefits but also residual risks remaining after adaptation, because adaptation rarely reduces impacts completely. This, in turn, influences the objectives set for adaptation (UNEP 2023), with the goal for such measures often being not to reduce climate risks completely but to manage them down to acceptable levels.

There is a strong economic case for many types of early adaptation action. The uncertainties of adaptation measures can make them difficult to justify when using standard economic frameworks, especially where adaptation involves up-front costs to deliver future benefits. However, there is a strong economic case for many types of early adaptation action. Adaptation planning may consider a range of complementary actions (Watkiss and Betts 2021), including

(i)   no- or low-regret adaptation actions that address existing and early economic costs of slow-onset changes and current extremes and so provide economic benefits today, such as from improved weather forecasts or seasonal predictions;

(ii)   climate-smart design in new infrastructure, where there is a cost-effective opportunity to help prepare for future impacts and/or avoid locking in large economic costs (Hallegatte et al. 2019); and

(iii)   low-cost preparatory and early actions that can be taken to improve future decisions and provide option value (DEFRA 2024), especially as part of adaptive management plans and adaptation pathways.

The economic case for moving beyond early action to transformational responses is strengthening, which requires preparation. As an example of actions that provide option value, easements that enable adaptation infrastructure, such as coastal protection measures, can be negotiated long before the infrastructure needs to be built. Other initial measures can be built to be compatible with varying levels of risk in the future, such as lower-height seawalls that have the ability to support height extensions. This is especially valuable for decisions that have long lead times or involve possible large-scale but uncertain impacts in the future.

3.1.2   Range of Adaptation Responses

Adaptation involves a wide range of actions with different characteristics. Climate adaptation can vary in terms of its intent and timing (Burton 2009) as shown in Figure 3.1. It can be reactive, responding to changes that are already being experienced, or proactive in anticipation of future effects of climate change. It may happen spontaneously (i.e., autonomous, unplanned, or not targeted specifically to climate risks) or be planned. Adaptation can also be incremental, seeking only to manage risks that impact the status quo, or be transformational and transitioning to a new socioeconomic system (IPCC 2014). It may be implemented at an individual project level or at a strategic or policy level. There are also three types of activities under climate adaptation (EIB 2022): (i) adapted activities, i.e., climate-proofing development activities where adaptation is a secondary objective; (ii) activities with shared objectives of development and adaptation; and (iii) activities where adaptation is the primary objective. Adaptation can involve interventions in nonmarket sectors and in public goods as well as in market sectors, which is important for financing.

Figure 3.1: Typology of Climate Change Adaptation Responses

Source: Adapted from Watkiss, P. 2022. Chapter 2. Evidence from the Literature. In *Barriers to Financing Adaptation Actions in the UK: An Evidence Report*. Climate Change Committee.

Adaptation is not just a set of technical solutions, as it can often depend on structural, institutional, ecological, and behavioral actions. The actions include capacity building and institutional strengthening; awareness raising and nonstructural actions ("soft options"); and technical, structural, or engineered actions ("hard options") (IPCC 2022a). Adaptation also requires enabling conditions, which include institutional arrangements, governance frameworks, and policy and legislation. Importantly, these types of actions are not alternatives to each other, and adaptation is more effective if they are combined together.

Cascading risks and interactions with other hazards can alter the effectiveness of adaptation actions. Compounding risks can occur when two events happen at once and these events combine to exacerbate impacts. For example, Ranger, Mahul, and Monasterolo (2022) estimate that the "compounding multiplier" of two climatic events happening at once can increase impacts by 50% compared to the sum of the impacts if the same two events happened independently. There is also the potential for consequential risks, e.g., droughts followed by floods, which can lead to greater impacts than if these events occurred at distinctly different times (Kruczkiewicz et al. 2022). Meanwhile, interdependencies within economic systems can lead to cascading risks. For example, electricity is critical to a wide range of sectors, notably communications networks, transport logistics, and water infrastructure. This means that damage to electricity infrastructure in one sector or locality can lead to economic losses in another (Hallegatte, Rentschler, and Rozenberg 2019). When setting adaptation policies, it is important to identify and address these potential risks through multihazard assessment and analysis of interactions.

Adaptation to climate change is part of a broader continuum with development. Climate change can exacerbate the need for development investment, such as funding for additional flood or coastal protection or for electricity for increased cooling in buildings. Conceptually, it can be considered to do so by shifting an existing demand (damage cost) curve for these interventions to the right, as weather-driven events such as flooding become more frequent and destructive under climate change (Figure 3.2). Addressing lower magnitudes of damage events (the left part of the demand curve) gives both existing and adaptation benefits, whereas addressing higher damages that are more unlikely without climate change gives a higher share of adaptation benefits. No-regret options exist up to the intersection of the existing damage and protection cost curves, whereas pure adaptation occurs only when that intersection is exceeded. Given that most developing countries have not reached the no-regret intersection for many protection measures, there are many no-regret opportunities that represent both adaptation needs and development needs that exist even without climate change. Thus, the distinction between what is development policy and what is adaptation policy is not always clear. Specifying the adaptation components of broader development initiatives is not always straightforward or appropriate, which complicates efforts to track adaptation progress and resource allocation.

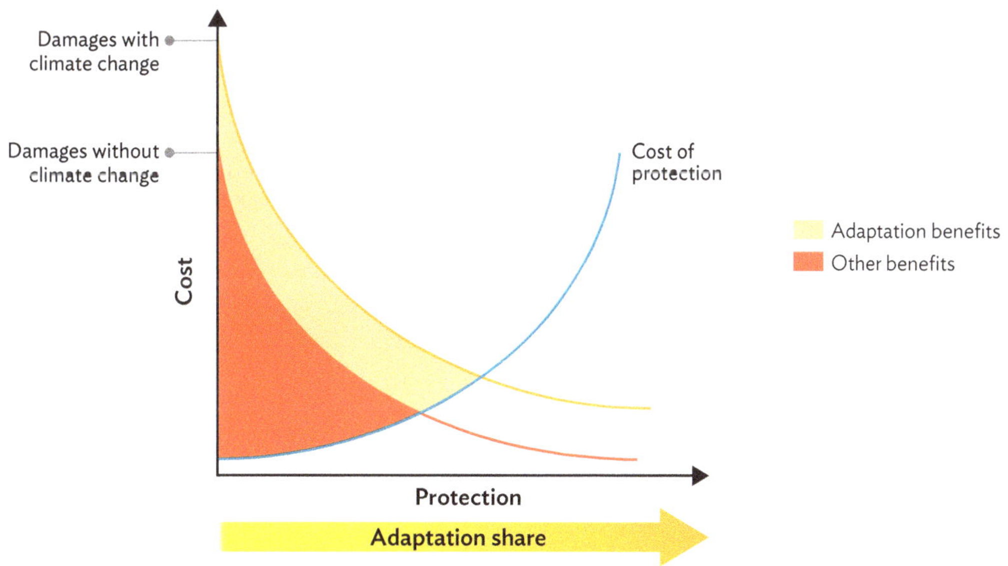

Source. Asian Development Bank visualization.

Trade-offs between adaptation as a direct goal and development as a means of adaptation need to be managed. Societies with higher human and social capital, as well as higher incomes and savings, are generally considered more resilient and have better adaptive capacity. Development options that improve and insulate livelihoods from weather can be considered a form of adaptation, especially for the least-developed economies (Dercon 2004). Indeed, there are potential trade-offs if resources are allocated to reduction of uncertain risks at the expense of growth that improves adaptive capacity. However, pursing development without considering climate change can lead to maladaptation, as development may be in sectors, activities, or areas with exposure to increasing damages. More explicitly developed climate-resilient development pathways (IPCC 2022a) that can deliver climate-compatible growth are often needed (OECD 2023b).

3.1.3   The Limits of Adaptation

Asia and the Pacific will increasingly face limits to climate adaptation. Due to economic and biophysical limitations, adaptation will not always be able to address climate impacts (IPCC 2022a). There can be "soft" limitations to adaptation. These include technical, financial, governance, and institutional issues that may be insufficient at present but can be overcome by addressing constraints. For example, low-lying coastal areas in certain islands of the Pacific might require investment in coastal protection that is 10 times a given location's gross domestic product in order to avoid catastrophic losses under sea-level rise, but that investment could be provided by those with historical responsibility for global warming. However, there are also "hard" limitations to adaptation where no adaptive actions are possible. For example, some natural ecosystems, such as warm-water coral colonies, are getting close to their biophysical threshold levels, and losses may be irreversible once those thresholds are exceeded (Dixon et al. 2022).

As the climate crisis deepens, there will be a need for more transformational adaptation. Once no-regret options are exhausted, pure adaptation investments are needed and, as increasing damage levels are confronted, they will need to be increasingly transformational. This not only involves technical responses but also takes account of the social, economic, and technological factors that can bring about rapid change at scale. For example, the scale of actions needed to address the impacts of sea-level rise on certain Pacific island countries will require different approaches to traditional coastal protection. This might include a combination of land-use planning and technical and nature-based solutions, but should also consider land reclamation, population consolidation in areas to be protected, migration, and governance changes (CRC 2020; UNF 2023).

Even if adaptation is rapidly scaled up, Asia and the Pacific will still face rising loss and damage. The concept of loss and damage (UNCC 2012) is concerned with the unavoidable and irreversible impacts of climate change, from slow-onset impacts as well as extremes, and includes both economic and noneconomic losses. It is driven by risks that will remain even after global mitigation action, as well as the residual risks if adaptation is not undertaken or is underdeployed. To cope, the Asia and Pacific region will need to scale up climate finance mechanisms to address the increasing frequency and intensity of extreme weather events. These include contingency financing, contingent disaster financing, as well as options such as sovereign risk pooling and new insurance approaches, nature-based insurance models, parametric insurance models, and innovative insurance instruments (Lloyds 2018).

3.2   Adaptation Needs and Opportunities

3.2.1   Sectoral and System Needs for Planned Adaptation

Adaptation is relevant to all sectors, although some sectors have more pronounced needs. Climate adaptation needs are, in many respects, specific to certain sectors and contexts, although they also often require multisectoral and systems-based planning and coordination to be effective. This section identifies some of the concrete needs for adaptation by sector, along with potential priorities for scaling up investment. The focus is on planned adaptation—recognizing that, while there will be autonomous and reactive responses, these on their own will not be sufficient to address the scale of present and future climate challenges. In addition, the focus is on measures that go beyond climate-proofing and target adaptation as a primary objective. The systems and sectors assessed include some of those most directly vulnerable to climate change, although adaptation needs extend into every sector.

Modeling reveals large investment needs for infrastructure, coastal and flood protection, and agriculture. UNEP (2023) compiles estimates from the Dynamic Interactive Vulnerability Assessment (DIVA) model for coastal infrastructure, the Global Flood Risk with Image Scenarios (GLOFRIS) model for riverine infrastructure, the IMPACT partial equilibrium agricultural sector model of the International Food Policy Research Institute, World Bank modeling of climate-proofing infrastructure, and more indicative modeling for other sectors in developing countries.[1] The estimates should not be considered exhaustive, as each model only considers a specific range of measures, and other possible options such as water retention basins or education investment are excluded. Nevertheless, the compilation is informative as a lower-bound set of estimates that is comparable across economies. Beyond the general category for infrastructure at $37 billion per year, coastal protection has the next largest investment need at $27 billion per year, followed by river flood protection at $17 billion and agriculture at about $7 billion (Figure 3.3). Agriculture and coastal protection dominate in the Pacific. Infrastructure and coastal and river flood protection dominate needs in East Asia, South Asia, and Southeast Asia. Meanwhile, investment in disaster risk reduction is largest for Central Asia. On a per capita basis, costs are highest in the Pacific, Southeast Asia, and East Asia.

Figure 3.3: Modeled Annual Adaptation Needs for Asia and the Pacific, 2023–2030

EWS = early warning system, SP = social protection.
Notes: Central Asia comprises Armenia, Azerbaijan, Georgia, Kazakhstan, the Kyrgyz Republic, Tajikistan, Turkmenistan, and Uzbekistan. East Asia includes the People's Republic of China and Mongolia. The Pacific comprises the Cook Islands, the Federated States of Micronesia, Fiji, Kiribati, the Marshall Islands, Nauru, Niue, Palau, Papua New Guinea, Samoa, Solomon Islands, Tonga, Tuvalu, and Vanuatu. South Asia comprises Afghanistan, Bangladesh, Bhutan, India, Maldives, Nepal, Pakistan, and Sri Lanka. Southeast Asia comprises Cambodia, Indonesia, the Lao People's Democratic Republic, Malaysia, Myanmar, the Philippines, Thailand, Timor-Leste, and Viet Nam.
Source: Authors' calculations from data generated by United Nations Environment Programme. 2023. *Adaptation Gap Report 2023: Underfinanced. Underprepared. Inadequate Investment and Planning on Climate Adaptation Leaves World Exposed*.

---

[1]   The modeling was performed under the Shared Socioeconomic Pathway 2, Representative Concentration Pathway 4.5 scenario (SSP2-4.5), which is similar to warming under full implementation and extension of nationally determined contributions submitted to the United Nations Framework Convention on Climate Change. Up to 2030, however, there is minimal divergence among Representative Concentration Pathway scenarios.

3.2.1.1 Coastal Systems

Extensive adaptation investment is needed to contain sea-level rise impacts in the Asia and Pacific region. The region faces some of the highest potential impacts of sea-level rise globally, due to its large coastal populations living in low-lying areas (112 million people in the region live in areas where the elevation is below 3 meters). When considered in the context of increasing storm surges and wave size under climate change, rising sea level has the potential to inflict very significant economic and social impacts, particularly for small island economies in the Pacific.

Adaptation to coastal risks, including sea-level rise, uses four strategies: protect, accommodate, retreat, and advance. Protection reduces coastal risks through hard engineered options (e.g., dikes for protection), soft options (e.g., sand replenishment to address erosion), or nature-based solutions (IPCC 2022b). Accommodation adjusts the use of the coastal zone in response to risks, such as with early warning systems, increasing insurance or increasing flood resilience (e.g., raising houses). Planned retreat involves pulling back from the coast via development control, land-use planning, or set-back zones. In practice, the strategies of protect, accommodate, and retreat have important synergistic roles to play in an integrated and sequenced response. Options to advance include land reclamation. The specific options (or mix of options) will vary with the coastal site and context, including current and future population levels and value of assets at risk.

Coastal adaptation solutions can have substantial net economic benefits. Addressing sea-level rise requires large investment in protection, with modeling studies estimating coastal adaptation costs at $27 billion per year until 2030 (UNEP 2023). However, this investment will generate high benefits. For example, other estimates of specific coastal adaptation costs among the developing member countries (DMCs) of the Asian Development Bank (ADB) range from $9 billion to $17 billion per year, but the net economic benefits are quantified to be at least 10 times the level of investment (Giardino and Vousdoukas 2024). Such investments also generally lead to wider macroeconomic and fiscal benefits (Parrado et al. 2020).

Ecosystem-based measures have potential for coastal protection. This includes restoring coastal ecosystems (e.g., mangroves) that can capture sediment and attenuate wave energy to reduce coastal flood risks. Such restoration also provides wider ecosystem service benefits to communities and the environment. Tiggeloven, Aerts, and Ward (2021) report that shoreline vegetation globally could contribute to important decreases in both absolute and relative flood risk (i.e., a 9%–13% reduction). Figure 3.4 shows the results of modeling on the global annual benefits from flood protection with and without mangroves. It indicates that mangroves provide flood protection benefits exceeding $65 billion per year in terms of avoided property losses. If mangroves were lost, an additional 40,000 square kilometers of land would be flooded across the world annually and 15 million more people would be exposed. Coral reefs are another example of how nature can provide adaptation benefits, providing natural coastal protection by reducing wave energy and buffering shorelines against soil erosion and storm damage. It is estimated that coral reefs can reduce wave energy by up to 97%, depending on the density and health of the reef (PCMSC 2022) and also provide wider ecosystem benefits, including food sources, employment and tourism opportunities, and recreation for local communities. Ecosystem-based solutions can also help address multiple risks from flooding, drought, storm surges, and sea-level rise. At the same time, these solutions are often most effective when used as part of a complementary portfolio of actions (Seddon et al. 2020; Chausson et al. 2020).

Figure 3.4: Global Annual Benefits Quantified from Flood Protection via Mangroves

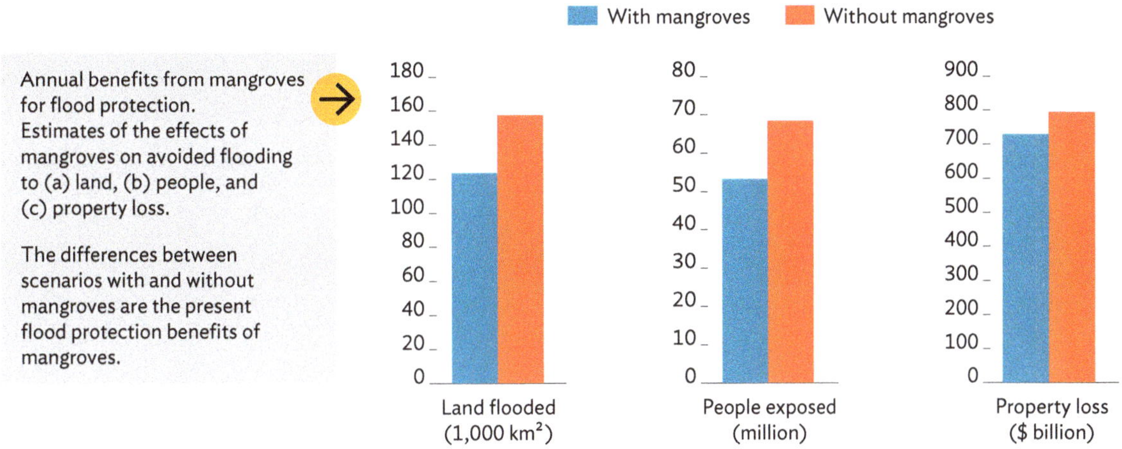

km² = square kilometer.
Source: Pelayo Menéndez, R., I. Monasterolo, F. Bosello, and S. Brown. 2020. The Global Flood Protection Benefits of Mangroves. *Scientific Reports*. 10 March.

Spatial planning and land-use policies are important for coastal protection. The mapping of zones facing potential inundation by current and future storm surges and sea-level rise can be used in land-use planning, including integrated coastal zone management planning, to anticipate and reduce current and future climate risks. Such mapping is particularly useful because of the generally irreversible nature of spatial planning decisions, which may otherwise lock in exposure to future risks. However, this level of planning depends on high-quality digital information such as precision topographical and elevation data from satellites as well as detailed inventories of critical infrastructure.

The vulnerability of coastal communities can be reduced by early warning systems. These systems are important to address increasingly frequent and intense cyclones and tropical storms and their associated storm surges and extreme waves, which are more destructive under sea-level rise. These systems are most effective if employed in combination with climate-resilient flood shelters, detailed emergency response plans, and accessible evacuation routes and protocols (Box 3.1 details reductions in death tolls achieved through the construction of flood shelters in Bangladesh). Preparedness for the risks associated with extreme weather events can be further enhanced by engaging local people in adaptation planning and decision-making processes, leveraging their traditional knowledge and localized expertise to develop context-specific adaptation measures and enhance community resilience.

Transformational adaptation will be needed to address sea-level rise in the Pacific. The up to 1 meter of mean sea-level rise that may happen within this century will pose large challenges for Pacific island countries (IPCC 2022b). Rises in sea level will vary regionally, due to ocean dynamics and land-ice loss (Oppenheimer et al. 2019), with sea-level rise in the South Pacific projected to be higher than the global average. This will mean greater adaptation needs and coastal protection costs that equate to significant proportions of gross domestic product. Adaptation costs are expected to be high for atolls, small islands, and low-lying deltas as these localities tend to have a very high percentage of the population living in low-lying areas. It is possible that there will be economic and social barriers to such expensive and long-term coastal protection strategies, especially under high emissions scenarios (IPCC 2022b). The solution is to find more strategic approaches to investment, including more transformational adaptation pathways. This is likely to involve policy support and capacity building for sustainable land-use planning and nature-based solutions; information and awareness

Box 3.1: Reducing Vulnerability via Flood Shelters in Bangladesh

Flood shelters, also known as cyclone shelters or flood-resistant shelters, play a crucial role in mitigating the impacts of flooding, cyclones, and storm surges in Bangladesh, a country prone to frequent and severe disasters triggered by natural hazards.

The construction of flood shelters in Bangladesh has dramatically reduced the number of fatalities from cyclones. For example, the 1970 Bhola cyclone is estimated to have claimed 300,000 to 500,000 lives, making it one of the deadliest storms in history. Subsequent cyclones continued to have high casualty rates, with about 11,000 deaths in 1985 and 140,000 in 1991.

Significant improvements in cyclone preparedness and the construction of shelters have led to a marked decrease in deaths from such disasters. Cyclone Sidr in 2007 resulted in approximately 3,500 deaths while Cyclone Aila in 2009 caused about 190 fatalities. More recent cyclones have seen even fewer deaths, illustrating the effectiveness of the shelters and other disaster preparedness measures. Cyclone Amphan in 2020 resulted in around 20 to 26 deaths.

This reduction in fatalities can largely be attributed to the increased number of flood shelters—from only 42 in 1970 to over 12,000 by 2020, capable of protecting nearly 5 million people. These shelters, coupled with improved forecasting, warning systems, and evacuation processes, have been critical in saving lives during extreme weather events.

Besides saving lives, flood shelters also contribute to community resilience by strengthening preparedness, response, and recovery capabilities in disaster-prone areas. By providing a centralized and well-equipped facility for emergency operations and coordination, flood shelters help communities mobilize resources, organize evacuations, and respond effectively to disasters. As such, they promote social cohesion and community solidarity by bringing people together during times of crisis, offering communal spaces where people can gather together to seek safety and support. This fosters a sense of solidarity, mutual assistance, and resilience, strengthening social networks and community bonds in times of adversity.

Overall, flood shelters play a vital role in Bangladesh's disaster risk management and resilience-building efforts, providing a lifeline for vulnerable communities and helping save lives, protect property, and build resilience in the face of recurrent natural hazards exacerbated by climate change.

Sources: Alam, E. 2024. Reasons for Non-Evacuation and Shelter-Seeking Behaviour of Local Population Following Cyclone Warnings Along the Bangladesh Coast. *Progress in Disaster Science.* 21; UCL (University College London). 2020. *Analysis: Bangladesh Has Saved Thousands of Lives from a Devastating Cyclone—Here's How.*

campaigns for the general public; new and innovative financing instruments; and strategic investments in gray, green, and hybrid projects. Such options might include considerations around major land reclamation (advance) or migration (an extreme form of retreat). These adaptation pathways should also address major social and economic issues and will depend on existing legal and institutional factors.

3.2.1.2 Water Management and Flood Control Systems

Changes to water availability and water-related stress are major channels of climate impacts. Water management encompasses a range of activities aimed at optimizing the use and conservation of water resources under climate variability and climate change, including the risks of floods and droughts.

Water Source Management

Balancing the demand, supply, and quality of water is critical under climate change. Reductions in water demand can be achieved by introducing climate-resilient agricultural practices; promoting water efficiency measures in households, businesses, and industry; reducing leakages within distribution networks (surface water and piped systems); and implementing incentive measures such as pricing of water and the introduction of water markets. Depending on location and physical context, water supply can be augmented by means of storage facilities (e.g., reservoirs, ponds, and wetlands); harvesting of rainwater; sustainable maintenance and use of groundwater reserves (e.g., by means of managed aquifer recharge and aquifer storage and retrieval); improved watershed management, which also helps combat floods and droughts; and enhanced systems for water transfer. The performance of water distribution and allocation systems can also be improved through the use of digital data sources and remote-sensing technologies. Water quality can be upgraded via increased sewage treatment, targeted land-use regulations (e.g., to reduce agricultural pollution of waterways), and expanded regulations on industrial pollution.

There are many no-regret options for managing water demand. Chief among these options are water efficiency measures, which have some of the highest cost–benefit ratios of all adaptation measures and should be a priority for early adaptation strategies. Water efficiency measures offer immediate benefits to users while also reducing the need for additional capital-intensive water supply projects (e.g., new storage reservoirs). As well as technological options, such measures include raising awareness of the benefits of water efficiency and the application of consistent pricing for all water users.

Integrated water resources management can coordinate the measures needed. Integrated river basin plans can be used to identify which measures are most suited to improve the performance of the water system under climate change. Integrated river basin management coordinates different forms of land use and water use and helps to manage competition for water between agricultural (the largest water consumer), industrial, residential, and ecosystem usage. The success of integrated water resources management critically depends on improving capacity to implement land-use regulations and spatial planning, as well as environmental regulations, to conserve water and land resources.

Floods

Many flood risks can be reduced by targeted adaptation actions. Chapter 2 of this report identifies the large flood risks arising from climate change in the Asia and Pacific region, along with the high economic costs associated with leaving these risks unresolved. Fortunately, there is a wide range of adaptation options for addressing inland (river and surface water) flood risks. The scope of flood risk management is determined by a combination of the hazard (the probability of the flood event), the exposure (what will be flooded) and the vulnerability (who and what might be impacted and by how much). Adaptation options to reduce the hazard are to store water locally and/or upstream (e.g., in reservoirs and detention basins) and through improved watershed management, riparian buffer zones, and wetland restoration. The hazard can also be reduced by flood protection infrastructure (levees, floodwalls, etc.) and river engineering works. The exposure can be reduced via floodplain planning and zoning (restricting or guiding development in flood-prone areas). Reducing the vulnerability can be achieved by implementing effective early warning systems (with community engagement) and promoting adaptive building design and retrofitting (e.g., elevated structures and flood-proofing).

The adaptation needs for flood risk management in Asia and the Pacific are large. Based on the global GLOFRIS model (Ward et al. 2017), the annual investment needed for adaptation measures to address river floods in ADB's DMCs is estimated at $17 billion per year through 2030 (UNEP 2023). However, these investments are expected to deliver very high net economic benefits (Ward et al. 2017). While hard engineered protection will be needed, the economic benefits of flood risk management can increase when soft options and nature-based solutions are included as part of a portfolio of adaptation actions. For example, the median benefit-to-cost ratio for flood prevention and preparedness is reported at 3:1 to 5:1 (World Bank 2021; Mechler 2016), but this ratio exceeds 5:1 (sometimes substantially so) when the flood risk management strategy also includes early warning systems. Box 3.2 provides some further context around the use of early warning systems in Asia and the Pacific.

Droughts

The key to reducing drought hazard is to create water supply surplus and redundancy. This can be achieved with a mix of "green" (nature based) and "gray" (constructed) infrastructure to enhance renewable water supply and lower demand through reductions in consumption. Supply can be increased through new storage and/or transfer of water from areas with abundant water (interbasin transfer). At local levels, rainwater can be harvested and better use can be made of groundwater resources. Managed aquifer recharge can replenish depleted groundwater resources during wet periods and sustain water availability during droughts (Box 3.3). Meanwhile, reducing the water demand of agriculture can be achieved by introducing volumetric water pricing and improved irrigation practices. General demand on water reserves can also be lowered by water recycling and reuse (e.g., from treated wastewater), by water pricing and providing financial incentives, and by public awareness and education on water-saving practices. Vulnerability to water stress can be reduced by implementing drought early warning systems to provide timely alerts and information to water managers, policymakers, and affected communities, enabling them to better manage scarce water resources and take proactive mitigation actions. Drought insurance and risk transfer instruments can also help to reduce the vulnerability of communities faced with water stress.

Water Supply, Sanitation, and Hygiene

Water supply, sanitation, and hygiene (WASH) services need adjustment to cope with increasing climate stresses. WASH infrastructure supplies drinking water, provides sanitation facilities, and treats wastewater. Under climate change, the provision of WASH services can be affected by the higher incidence of floods (physical damage to infrastructure, health risks) and more prolonged droughts (reduced surface water and groundwater supplies). As a result, there is a need to ensure climate resilience is integrated into WASH investments (Hallegatte et al. 2019). Protecting infrastructure includes, for example, moving water intake points away from areas of known flood risk and adapting the design and construction of processing plants to reduce vulnerability to hazards such as storm surges. Attention may be given to the threats to water quality from increased rainfall and floods. In addition, improved water supply can prevent transmission of mosquito-borne diseases expected to increase under climate change by reducing breeding sites from informal water storage, as illustrated in Box 3.4.

Box 3.3: Increasing Groundwater Availability by Managed Aquifer Recharge in Mongolia

The limited and very seasonal rainfall in Mongolia makes groundwater an important source of water for drinking purposes and industrial use, particularly in mining. With demands on groundwater increasing, capturing surface river flows to recharge groundwater aquifers may be a viable water storage and delivery option. To this end, the Asian Development Bank assisted the Government of Mongolia in developing a managed aquifer recharge (MAR) pilot project to test and evaluate the applicability of this technology.

This pilot project was conducted in Baganuur and comprised a series of shallow swales (channels) traversing a river known as the Nariin Stream. The swales are designed to intercept excess river and overland flows resulting from heavy rainfall. As the wetted area expands, surface water filters down into the subsurface. Infiltration recharge is accelerated through a series of shallow holes drilled into the swale beds, which are in-filled with gravel and lined with a geotextile filter.

While the MAR pilot successfully demonstrated the potential for sustainable groundwater replenishment in Mongolia, further monitoring and assessment are required. However, based on an assessment that considered the available geological and hydrogeological data, surface water resources, and localized user demands—several sites have been proposed for possible replication of the MAR pilot for domestic and mine water use.

Source: Fan, M. 2023. Managed Aquifer Recharge in Mongolia: Policy Recommendations and Lessons Learned from Pilot Applications. ADB Briefs. No. 263. Asian Development Bank.

Box 3.4: Reducing Disease Through Climate-Resilient Municipal Services in Nepal

Investments in water supply, sanitation, and hygiene (WASH) can help reduce health risks exacerbated by climate change. The Sixth Assessment Report of the United Nations Intergovernmental Panel on Climate Change projects that the risk of vector-borne diseases (e.g., malaria, dengue) will increase if no climate change adaptation measures are taken by individual economies. Due to increased temperatures and altered weather patterns, more localities will become suitable breeding grounds for vectors of disease, such as mosquitoes.

Based on a 2023 study supported by the Asian Development Bank, without appropriate action on WASH services, the number of malaria cases in Nepal could increase by up to 27% over 2011–2040, with the number of dengue cases predicted to increase by up to 44% over the same period. Over 2041–2070, case numbers could rise by as much as 69% for malaria and 113% for dengue. However, as shown in the figure below, better access to drinking (tap) water and improved solid waste management at the household level have the potential to reduce disease risks by eliminating informal water storage that serves as mosquito breeding sites.

Modeled Dengue Cases With and Without Investments in Urban Water Supply and Solid Waste Management at the Household Level

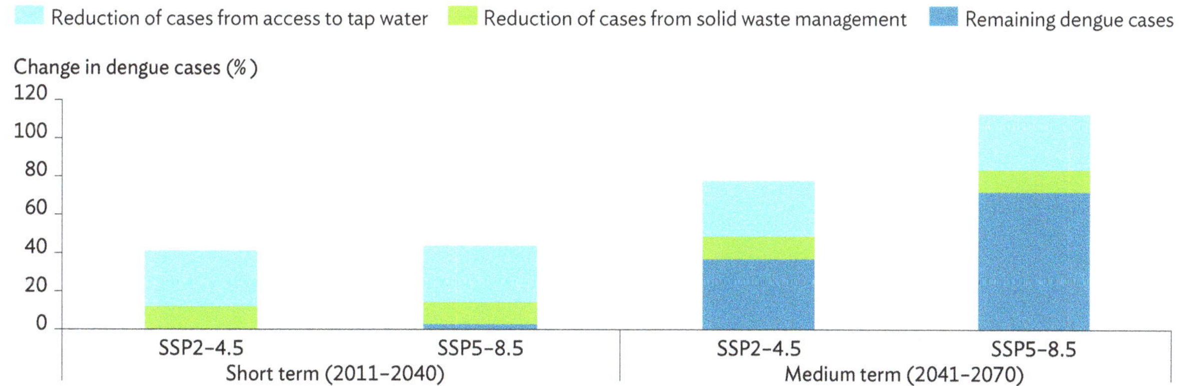

SSP = shared socioeconomic pathway.
Source: Boelee, E., A. Giardino, and A. Conroy. 2023. Investments in WASH Can Help Nepal Reduce Climate Health Risks. Development Asia. 16 January.

continued on next page

Box 3.4 *continued*

The study also showed that Nepal's investments to reach the Sustainable Development Goals for water and sanitation and improving waste management can reduce malaria by 17% and dengue by 41% in urban areas. Improving WASH services can also deliver other health benefits, such as lessening diarrheal disease and increasing good hygiene practices, as well as helping to build resilience to vector-borne diseases.

Source: Boelee, E., A. Giardino, and A. Conroy. 2023. Investments in WASH Can Help Nepal Reduce Climate Health Risks. *Development Asia*. 16 January.

3.2.1.3 Agriculture and Food Systems

Agriculture has many no-regret opportunities for climate adaptation. Such opportunities can reduce existing weather- and climate-related impacts and build the sector's resilience to future effects of climate change. Agriculture and food systems will also be strongly affected by mitigation (Box 3.5), so efforts to improve productivity are essential, whether to adapt to pressures of climate change from more warming or from competition for land, water, and energy under efforts to reduce emissions.

Improved weather and climate information services are early no-regret adaptation options. Weather and climate variability already cause substantial losses to farmers and agricultural enterprises in Asia and the Pacific, as shown in Chapter 2. These impacts can be reduced through improved or new weather services, such as shorter-term forecasts and early warning systems for extreme weather events. Impacts can also be eased by enhanced climate information services, including seasonal and longer-term forecasting. These services can be tailored to agriculture or water management (including irrigation) and may deliver substantial economic benefits by avoiding losses and/or improving yields (Rouillard et al. 2016). There is potential for innovation in advanced technology in these areas, such as the use of artificial intelligence and real-time earth observation data for early warning systems (ADB 2023a) as well as innovative financial products, including index-based crop insurance.

Precision agriculture provides opportunities for climate-smart technologies to enhance resilience. Precision agriculture involves the use of information and technology to enable site-specific management practices that optimize responses to climate, weather, and site information. It can help adjust practices to varying climatic conditions and improve use of scant resources, such as water, so as to address productivity variations across time and space (Roy and George 2020). As agriculture becomes increasingly mechanized in Asia and the Pacific, precision agriculture technology can be embedded into new machinery. More broadly, mechanization and reduced tillage can help reduce the need for human labor in periods when high temperatures restrict the capacity to work. Integrated pest management can also offer more resilient and sustainable control of new pests and diseases (Babendrier et al. 2019). Other agronomic practices, such as agroforestry, intercropping, and cover crops, have potential to improve climate resilience but often need complementary technologies, such as appropriate machinery, to achieve compatibility with farmer labor and other constraints, as adoption of these practices has been slow (Stevenson et al. 2019).

Box 3.5: The Impacts of Emissions Mitigation on Agriculture

Many of the same socioeconomic systems are subject both to the impacts of climate change and to subsequent actions to mitigate climate change by reducing emissions of greenhouse gases. As a result, these systems need adaptation both to climate change and the pressures created by emissions mitigation. A primary example is agriculture, which is vulnerable to increased frequency of extreme climatic events due to climate change. However, achieving goals under the Paris Agreement on Climate Change is likely to depend on expanding carbon sequestration through forests, which limits the availability of land for agriculture. In most scenarios of Paris Agreement compatible emissions pathways, bioenergy use rises, often in combination with carbon capture and storage, and bioenergy production also comes at the expense of land area available for growing food crops.

In 2023, the Asian Development Bank released a study (ADB 2023) that models a set of scenarios compatible with the Paris Agreement, then contrasts these scenarios with current policies. This modeling considers both the deployment of bioenergy and use of expanded forest cover, in addition to other energy-related mitigation options. It finds that the land area for growing food crops in developing Asia decreases by 36 million hectares by 2050 and 58 million hectares by 2070 in the Paris Agreement compatible scenarios (as indicated in the first figure below). Much of this is driven by expansion of bioenergy crops, which reach 97 million hectares in the region by 2070. As a result, food prices reach values that are 35% higher than they would be under current policies by 2065 (as indicated in the second figure below). Agriculture therefore clearly needs adaptation to emissions mitigation policies, as well as to the impacts of climate change, to avoid adverse food security outcomes.

Land-Use Patterns Under Current Policies and Under Emissions Scenarios Aligned with the Paris Agreement

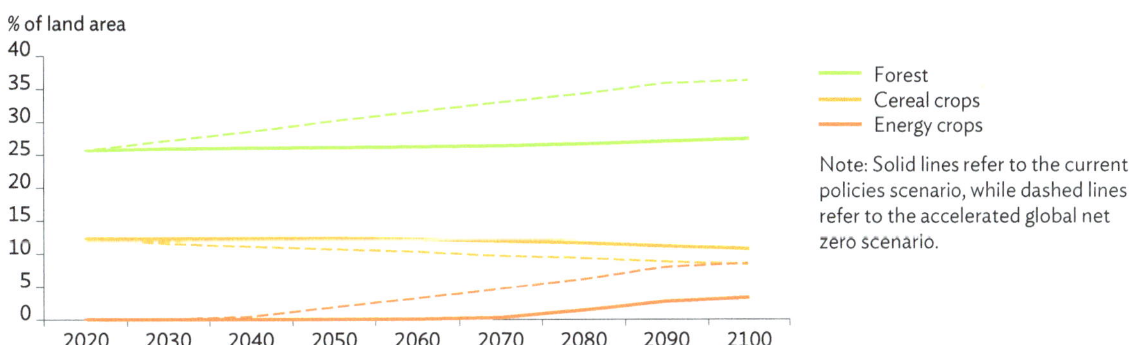

Source: Asian Development Bank (ADB). 2023. *Asia in the Global Transition to Net Zero: Asian Development Outlook 2023 Thematic Report.*

Food Prices Under Current Policies and Under Emissions Scenarios Aligned with the Paris Agreement

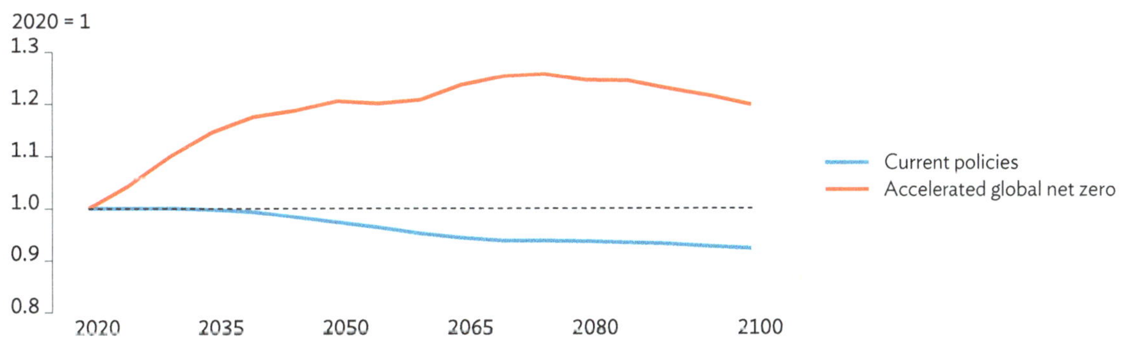

Source: ADB. 2023. *Asia in the Global Transition to Net Zero: Asian Development Outlook 2023 Thematic Report.*

Reference:
ADB. 2023. *Asia in the Global Transition to Net Zero: Asian Development Outlook 2023 Thematic Report.*

Source: Authors.

Embodied agricultural innovations, including stress-tolerant varieties, can help address climate shifts using existing extension systems. Sulser et al. (2021) identify that agricultural research has a key role for adaptation and related studies provide very positive cost–benefit ratios for investing in research to address climate change risks (>30:1) (Rosegrant et al. 2023). High-returns research products include stress-tolerant crop varieties, especially for drought and floods. Increasingly, the private sector is investing in developing varieties adapted to changing climates (Smyth, Webb, and Phillips 2021). Existing agricultural extension systems have demonstrated track records of getting farmers to adopt improved varieties, especially for staple crops (Garbero, Marion, and Brailovskaya 2022). This success is facilitated by the fact that these innovations are embodied technologies, which are technologies that are packaged in ready-to-use tangible products, such as seeds that embody genetic innovations from research. These embodied innovations are directly transmitted as varieties spread via markets that often effectively reach farmers.

Enhanced extension capacity and complementary services are needed to spread knowledge-intensive adaptation technologies. Many adaptation options in agriculture—beyond improved crop varieties and machinery—rely on uptake of disembodied knowledge (not packaged in products), such as management and diversification advice. They thus require transmission of much more information among farmers and have diffused more slowly from existing extension systems (Stevenson et al. 2019). More iterative and interactive extension approaches, such as farmer field schools, demand responsive approaches, such as farmer contracted extension, and gender-sensitive approaches all show promise to improve the adoption of knowledge-intensive practices (Waddington et al. 2014; Mullen, Gray, and de Meyer 2015; Watkiss and Climato 2016). Incentives, such as productive asset transfers, are often also needed to overcome barriers (Phadera et al. 2019). Delivering these innovations will require capacity building, including policy strengthening and the mainstreaming of climate concerns into agricultural development and planning (Figure 3.5).

Figure 3.5: Climate Change Adaptation Responses and Supporting Actions for Agriculture

Water management	Effective water accounting, monitoring, and infrastructure are crucial, especially in drought-prone regions.
Agro-weather forecasting	Accurate climate information aids farmers in decision-making.
Climate-smart practices	Techniques such as soil and water management enhance resilience to climate extremes.
Crop diversification	Diverse crops mitigate risks and improve farmer incomes.
Access to finance	Tailored financial products provide a safety net against adverse weather impacts.
Knowledge sharing	Training programs build adaptive capacity among farmers.
Infrastructure resilience	Climate-proof infrastructure ensures market access for agricultural products.
Policy support	Supportive policies and incentives promote adoption of sustainable practices.

Source: Asian Development Bank construction.

Achieving better water-use efficiency will be critical as areas become more water-stressed. Agriculture is responsible for 90% of freshwater use in Asia and the Pacific, and the sector relies on some of the world's largest river basins, such as the Mekong and the Ganges, to support the livelihoods of hundreds of millions of people. These systems are already under large stress due to overextraction of water, salinity intrusion, and pollution, and climate change can act as a stress multiplier. These stresses are often driven by agricultural activities. As a result, the agriculture sector finds itself in a complex relationship with other water users, environmental values, conservation efforts, and the provision of ecosystem services. Integration of agricultural water use within integrated water resources management presents an opportunity to harmonize the use of water resources with cross-sectoral and environmental priorities. This can be complemented with measures that increase water productivity (e.g., crop diversification, water-saving irrigation techniques), volumetric water pricing, and adaptive allocation, especially in regions prone to droughts or water scarcity. Existing irrigation infrastructure will often need to be upgraded to support these measures, as water flows may need closer regulation and end uses may need metering. Maintaining and managing this enhanced infrastructure will often require institutional innovations, as maintenance has been a frequent challenge for irrigation systems.

The resilience of value chains for agriculture and food depends on a wide range of infrastructure. Climate change will not only affect agricultural production but also a range of systems along the value chain. Associated infrastructure for crop and food transportation, storage, processing, distribution, and trade will also face climate risks. In addition, destabilized production under a more volatile climate will lead to greater need for buffer stocks and storage to stabilize food supply and prices. Warmer temperatures will increase food spoilage and demand for cold storage (Duchenne-Moutien and Neetoo 2021). There also is emerging evidence that carbon dioxide fertilization of crops may depress micronutrient concentrations, which means that expanded biofortification in processing facilities and cold storage of more nutrient-rich foods may be needed to compensate (Ziska 2022).

Other anthropogenic stresses must be reduced for fisheries to cope with climate change. About one quarter of animal protein intake in Asia is from fish, with an even higher share in the Pacific. Climate change will profoundly affect natural fisheries resources, which are already under strong pressure from overextraction, pollution, and destruction of habitats. While effects of temperature rises and ocean acidification can only be addressed through climate change mitigation, these other pressures can be reduced to avoid compounding effects. Overextraction can be addressed by enforcing restrictions on fish-catch levels and on damaging fishing practices. Pollution can be addressed through better land-use planning, waste management, and enforcement of effluent standards. Habitats can be preserved through expansion and protection of high conservation value sites. As climate change will shift the distribution and population cycles of fish species, regulation will need to become more adaptive. Pressure on natural fisheries can also be reduced through expansion of aquaculture, although measures are needed to ensure that aquaculture is sustainable and does not introduce new pressures, such as introduction of exotic species. Many of these measures depend, in turn, on strengthened capacity for planning, regulation, and enforcement (Shelton 2014).

3.2.1.4   Energy Systems

Escalating energy demand for cooling systems can be anticipated. As highlighted in Chapter 2 of this report, climate change will be a major driver of energy demand in Asia and the Pacific, increasing demand for cooling among residences and businesses, with high economic costs. Without action, this response to rising temperatures will drive potentially substantial increases in greenhouse gas emissions. Planned adaptation measures can address these effects by, for example, increasing end-use efficiency (to best available levels, which might moderate annual electricity consumption increases by 40%) and incentivizing behavioral changes in cooling (with a mix of ventilation, fans, and air-conditioning), which can significantly reduce electricity use (Colelli, Wing, and Cian 2023). There are also opportunities to reduce the use of air-conditioning through passive cooling embedded in building designs (and building codes) as well as through green infrastructure options and options around urban planning and zoning.

There is potential to build climate resilience into energy supply. The Asia and Pacific region is receiving some of the largest flows of investment into energy supply, with an increasing share of this investment being in green and renewable energy. However, this new infrastructure is vulnerable to climate change, especially given its long lifetime, and it is important to address these climate risks during design, a concept known as "climate-proofing". Doing so involves low additional costs and is highly cost-effective, generating benefits four times greater than the costs (Hallegatte, Rentschler, and Rozenberg 2019). However, effective climate-proofing often requires enabling actions by government, including changes to design standards.

3.2.1.5 Social Protection Systems

Social assistance and social protection will need to be scaled up to protect the most vulnerable. Existing social assistance and protection programs include a range of interventions, such as cash transfers and public works. While these programs—and the people they seek to protect—will be affected by climate change, they also provide an opportunity to deliver support to the most vulnerable people in advance of extreme events and to help these people deal with the impacts of climate-related shocks. Social assistance and social protection programs have high economic cost–benefit ratios, up to 6:1 (Venton and Coulter 2013; Bangalore et al. 2016), but they need adjustment to respond to changing extremes and shocks. As women will be disproportionately affected by climate change, many social protection measures should be targeted to female beneficiaries.

Adaptive social protection systems can improve resilience to climate change. Social protection systems can include additional elements to integrate adaptive and event-responsive features. This includes targeting beneficiaries at risk, adaptive features to enable horizontal expansion (including additional beneficiaries) and vertical expansion (higher benefit amounts), and event-responsive contingency funds linked to early warning systems. The latter option can, for example, provide cash transfers to vulnerable communities in advance of a forecast climate-related hazard, such as a drought. Such initiatives are projected to have high economic benefits (DFID 2015), especially in highly vulnerable areas. Adaptive social protection also includes support to build shock-responsive systems, plans, and partnerships in localities that have high vulnerability, improving preparedness and allowing these systems to be scaled up after a large climate-related shock occurs. Social protection programs can also integrate measures to promote climate-resilient jobs and livelihood options along with skills in adaptation-related industries.

3.2.1.6 Health Systems

Many health impacts from climate change can be avoided in the short term. This can largely be achieved through increased deployment of preventative measures, with several reviews identifying a wealth of health adaptation options (Berrang-Ford et al. 2021; Scheelbeek et al. 2021). For existing health burdens that may be exacerbated by climate change (such as water- and food-borne disease, malaria), there are cost-effective prevention measures, as identified in the Disease Control Priorities (DCP3 n.d.). These measures can include more action to address increasing health burdens, such as climate-related water- and food-borne disease, especially among infants (Horton 2017) or cost-effective preventative measures for malaria (Ralaidovy et al. 2021). However, there will also be a need to consider how climate change is changing the geographic onus of health risks, with the potential need to transfer existing actions to new locations and to scale up or expand the reach of pilot programs such as health-related early warning systems.

Climate change will also increase the prevalence and range of new diseases. Climate change will lead to new disease burdens, including new and emerging climate-related vector-borne diseases (e.g., chikungunya, dengue). In response, the G20 High Level Independent Panel (G20 HLIP 2020) calls for more robust disease surveillance and detection networks, coupled with coordinated regional action and information sharing. This will require a major expansion of surveillance networks and the development of expertise at the global, regional, and national levels (since enhanced surveillance is a global and regional public good). This improved surveillance can be complemented by more-effective national early responses to disease outbreaks and by cost-effective preventative measures (Brady et al. 2020).

Addressing heat-related mortality in Asia and the Pacific requires a multifaceted approach. There are early adaptation options, based around heat-alert schemes and supportive health sector responses, which can be scaled up cost-effectively to address heat-related mortality (Ebi et al. 2004; Hunt et al. 2016). However, even with these options in place, additional actions will be needed, especially to assist the groups in society that are most at risk, including women. Addressing heat-related mortality also has important cross-linkages to the challenge of providing sustainable cooling systems through adaptive building design, nature-based solutions, and urban planning.

Adapting to the impacts of heat will be critical for maintaining labor productivity. Adaptation responses to the effects of increased heat will range from measures to offset reductions in labor output (e.g., changes in working practices, moving activities to different times of day) through to the protection of workers from heat stress by initiating heat awareness campaigns, implementing heat and health alerts, investing in cooling, and introducing new occupational health legislation, as illustrated by heat stress legislation in Singapore (MOM 2024). As women are at particular risk from heat stress, gender-sensitive approaches will be needed. Heat health plans are a first step to help coordinate responses.

Health systems and health services can broadly embed adaptation considerations. The World Health Organization (WHO 2015) recommends moving beyond individual health burdens to a more systematic approach that integrates climate adaptation into health programs and delivery, emergency preparedness and health information systems, health-related supply chains, and infrastructure such as hospitals and health facilities (including through retrofits and building new infrastructure).

3.2.1.7 Ecosystem Services

Planned adaptation is needed to support ecosystems and natural capital to adapt. This goes beyond deploying nature-based solutions for direct protection of human systems. Biodiversity and natural habitats (natural capital) provide key ecosystem services across Asia and the Pacific, but they will also need support to adapt to climate change, especially in the context of potentially irreversible impacts. This support involves actions that directly target protected and valuable areas with planned adaptation. It can include habitat conservation, rehabilitation, and restoration; more refugia (protective habitat for vulnerable and endangered species); enhanced connectivity, including through wildlife corridors; and even translocation. Such actions link to the global goal to halt and reverse nature loss by 2030 agreed at the 2022 United Nations Biodiversity Conference.

3.2.2 Cross-Sectoral Measures

3.2.2.1 Trade Policy

International trade and trade policy can be important in facilitating climate adaptation. International trade can help build resilience to climate-related shocks, improve food security, and provide access to technologies for climate adaptation. Trade is also likely to be strongly affected by climate change, with climate change impacting upon comparative advantages through productivity losses and supply shortages and increasing trade costs due to trade disruptions.

Trade is important in enabling access to adaptation technologies. Green technologies accounted for around 13% of all inventions (measured using patent data) over 2000–2022, with adaptation technologies accounting for around 10% of those green technologies (ADB 2024). Most green adaptation technologies are found in agriculture (33%), health (26%), water (16%), and infrastructure (14%). More than 90% of green technologies are invented in just five countries—the People's Republic of China (PRC), Germany, Japan, the Republic of Korea, and the United States—a concentration that is somewhat higher than for all technologies. Trade and global value chains therefore play an important role in the diffusion of adaptation technologies. Indeed, a strong association is found between countries trading bilaterally (in both traditional and value chain trade) and countries sharing patent protection for adaptation technologies, i.e., relationships between inventing countries and the countries in which inventors expect their inventions to be used.

Climate adaptation can be more explicitly integrated into trade policy. The importance of trade for climate adaptation has been reflected in the increased attention given to adaptation in plurilateral and multilateral trade policy developments. Since 2015, there has been an increase in the number of regional trade agreements that include provisions related to disasters triggered by natural hazards, including particular provisions relating to increased cooperation, exemptions, disaster management, the compensation of losses, and emergency trade facilitation (WTO 2022). There has also been an increase in the number of trade-related measures on adaptation notified to the World Trade Organization. These tend to be limited to certain sectors (agriculture, energy, forestry, manufacturing, and services) and include both support measures (e.g., grants and direct payments, nonmonetary support, loans, and financing) and other measures (e.g., technical regulations). Aid for trade disbursements for climate adaptation have also increased since 2015.

3.2.2.2 Domestic Markets

Poorly defined markets and property rights impede adaptive responses. Land markets, especially those for agricultural land, are subject to many distortions in Asia and the Pacific. Land tenure may be undocumented, incomplete, or tied to specific forms of land use. Other land is common property or subject to competing tenure claims. This impedes the ability of households and other market actors to respond to climate stresses by shifting land usage patterns, consolidating cultivated lands to enable mechanization, or pursuing alternative livelihoods. Similarly, many fisheries are effectively common resources for which management rules essential to climate resilience are difficult to enforce. Reforms to improve and create incentives for stewardship can enhance resilience.

Some policies obscure climate signals, which reduces autonomous adaptation. Markets often need reform to transmit or enable adaptive responses to climate change. For example, if water is scarce, high prices for water consumption encourage more-efficient use. Similarly, if hydropower storage is low, high electricity prices encourage more-efficient use. However, governments often interfere with or do not enable this transmission. For example, water is not charged based on usage and electricity prices are subsidized when production prices spike. Reforming these prices to transmit climate signals can increase autonomous adaptation.

Other policies can inadvertently encourage maladaptation. For example, many governments set minimum support prices for crops, which encourage production of crops that are climate-vulnerable, water-consumptive, or from which farmers would otherwise diversify. As another example, excessive disaster relief can cause households to insufficiently consider climate risks in housing and investment decisions. Similarly, expectations of flood or coastal protection infrastructure can create adverse anticipation effects in which those risks are insufficiently considered.

Innovative insurance products are needed to complement adaptation measures. Insurance has an important role in risk transfer (Mechler 2016), a form of climate adaptation, and has a key role in managing the residual risk of extremes. This can include mechanisms ranging from household insurance right up to national (sovereign) risk-pooling facilities (CCRIF 2010). However, with increasing extreme weather events, climate change will create issues around the affordability and/or coverage of such events. Moreover, traditional forms of insurance are not available to cover the slow-onset impacts of climate change. This calls for innovative approaches, with a range of new insurance instruments (Lloyds 2018), such as insurance-linked loan packages.

3.2.2.3 Households

Improved human capital has a key role in adaptation. Education has a key role in delivering climate adaptation. In general terms, education enhances resilience, in that more educated people have better understanding and management of risks (Raghupathi and Raghupathi 2020). More educated individuals also tend to have more livelihood opportunities, which enhances their resilience. They also are more likely to adopt innovations that help to reduce vulnerability. At the same time, the poorest and most climate-vulnerable countries tend to have the poorest-performing educational systems (OECD 2023a). Making education more effective is a key way to help enable climate adaptation and household resilience. Education can also enable adaptation directly by embedding climate change awareness and responses in curriculum materials from primary to tertiary level. Education can also build skills and capacity for employment in adaptation goods and services.

Greater access to finance can help improve coping ability and resilience. Savings, remittances, and access to finance can help households and businesses cope with climate-related shocks and adapt to climate change (Bangalore et al. 2016). They can help to smooth consumption when shocks arise or to help in recovery. Financial services, such as credit, can help households and businesses invest in adaptation measures and other opportunities to counter climate change. Financial inclusion is therefore vital in this process. Policies to improve financial inclusion include strengthening financial systems; enhancing financial infrastructure (including digital and mobile banking, which can allow faster payments and provide rapid access to remittances when shocks arise); and creating the enabling conditions for the private sector to offer new or combined financial instruments for resilience enhancing investments.

3.3   Adaptation Policy Responses to Date

3.3.1   Growing Policy Recognition of Adaptation

An increasing number of international processes target adaptation. The United Nations Framework Convention on Climate Change (UNFCCC) has developed workstreams and processes to support governments in identifying their country-level adaptation needs through National Adaptation Plans (NAPs). These plans identify both medium- and long-term adaptation needs and identify strategies and programs to address those needs (UNCC n.d.). The Global Goal on Adaptation was established under the Paris Agreement and, in 2023, Parties to the agreement adopted the UAE Framework for Global Climate Resilience, which includes a range of climate adaptation targets. A work program has since been established to develop indicators for measuring progress toward the targets outlined in the framework.

All Parties in Asia and the Pacific have communicated adaptation priorities to the UNFCCC, and many have identified financing needs. These priorities have been submitted in national reports to the UNFCCC, either in their NAPs, their nationally determined contributions (NDCs), or their National Communications on Climate Change (NCCCs). Many economies have also costed these priorities and estimated their adaptation financing needs. Of ADB's 41 DMCs, 31 have developed NAPs or have NAPs under development. Of the 10 DMCs without explicit NAPs, seven have published either a national adaptation plan of action or a joint national plan of action. The key difference between a "national plan of action" and a full NAP lies in the focus and time frame: the former addresses short-term adaptation needs and priorities, while the latter adopts a more comprehensive medium- and long-term perspective on national adaptation requirements.

Adaptation plans reveal a range of sector priorities. Figure 3.6 outlines an analysis of climate adaptation priorities for ADB's DMCs, based on a review of submitted NAPs, NDCs, and NCCCs in line with UNFCCC categories of adaptation. In this context, the top four adaptation priorities are agriculture, food production and nutrition security, terrestrial and wetland ecosystems, and water resources—all of which appear in over 75% of all submissions. A majority of DMCs also include adaptation priorities around human health (73%), biodiversity and ecosystems (73%), coastal and low-lying areas (66%), disaster risk management (61%), forests (59%), and ocean ecosystems (54%). However, while this indicates general priorities, it is important to note that national reports differ in their level of detail and plans differ in terms of their adaptation ambitions, socioeconomic circumstances, consideration of future climate scenarios, methods employed to identify and prioritize adaptation options, sectoral coverage, and implementation time frames (UNFCCC 2022).

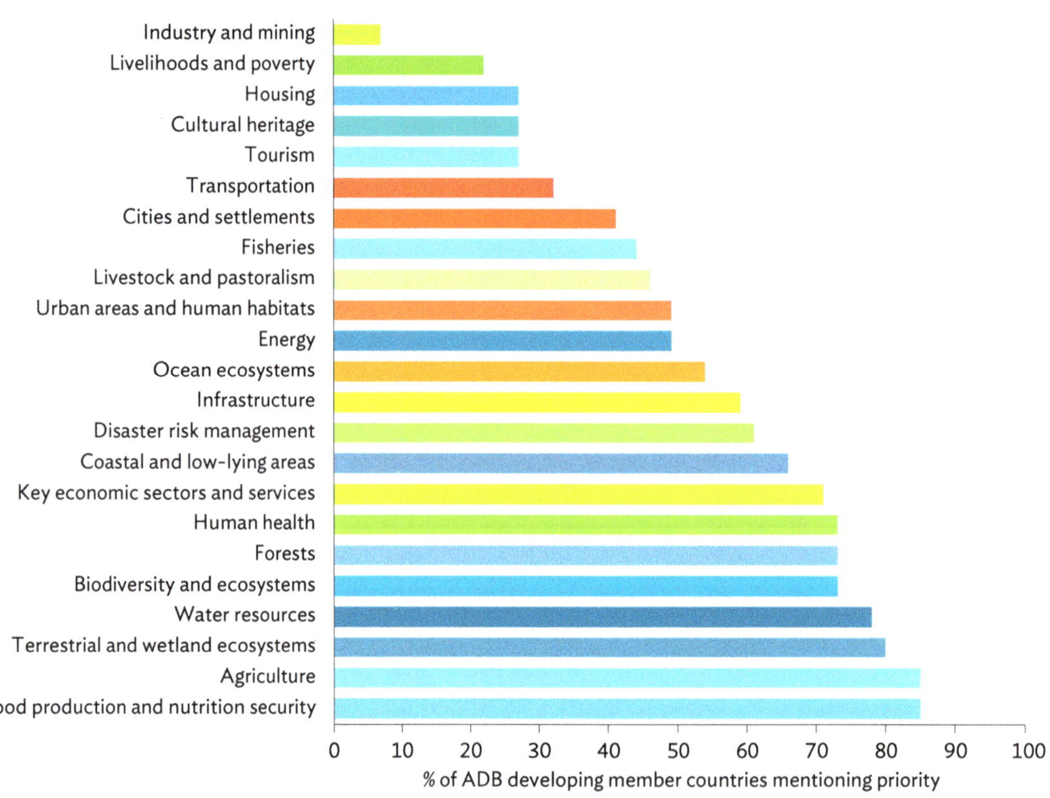

Figure 3.6: Climate Adaptation Priorities for ADB Developing Member Countries

Notes: Percentages denote the proportion of countries mentioning each priority in National Adaptation Plans, nationally determined contributions, and/or National Communications on Climate Change submitted to the United Nations Framework Convention on Climate Change. Data cover 41 Asian Development Bank developing member economies, excluding Brunei Darussalam; Hong Kong, China; the Republic of Korea; Singapore; and Taipei,China.
Source: Markandya, A., P. Orgen, and S. Abdulla. 2024. Adaptation Responses to Date in Asia-Pacific Region. Background paper for the *Asia–Pacific Climate Report 2024: Catalyzing Finance and Policy Solutions.* Asian Development Bank.

Systematic prioritization could be improved by wider use of economic analysis. Many national adaptation documents involve long lists of stand-alone adaptation priorities, with associated financial needs that are large. More recently, countries have started to integrate adaptation priorities into national and sector development planning processes, linking climate adaptation with national developmental goals. However, while several DMCs have reported financial costs in their NAPs, NDCs, and NCCCs (Orgen and Abdulla 2024), almost none of them identify the quantified benefits of adaptation (UNFCCC 2022). Moreover, not all DMCs quantify adaptation investment needs on a sectoral basis, which suggests that the costs of specific options in each sector are not always assessed systematically.

Countries can make use of more policy instruments to scale up adaptation. An analysis of policy instruments employed by ADB's DMCs to support adaptation has been compiled for this report, based on the Climate-Laws database as of May 2024 . As shown in Figure 3.7, of ADB's 41 DMCs, 33 had implemented policy actions under disclosure obligations and standards that can help raise awareness and action on risks. Almost all the DMCs assessed (40 of 41) had issued or disclosed national policy statements and strategy documents highlighting efforts toward international cooperation, resilience targets, long-term plans, and financing NAPs. These efforts primarily fall under national planning strategies and are increasingly linked to national development plans. Close to half the DMCs (19 of 41) had provided economic subsidies, while 22 had offered direct provisions for climate funding and investments to support adaptation efforts, and 16 had offered tax incentives. Examples of subsidized activities include solar irrigation pumps in drought-prone areas of India and water management, embankment improvements, and cyclonic preparedness measures in Bangladesh. As an example of tax incentives, the PRC offers tax rebates for enterprises involved in water conservation and management, particularly in water-stressed localities, to promote sustainable water use. Five DMCs had used carbon pricing policies to reallocate public revenue toward adaptation. The great diversity of policy instruments applied across DMCs —combined with the fact that no instruments are used in most of these DMCs— indicates that there are many instruments still available to strengthen adaptation.

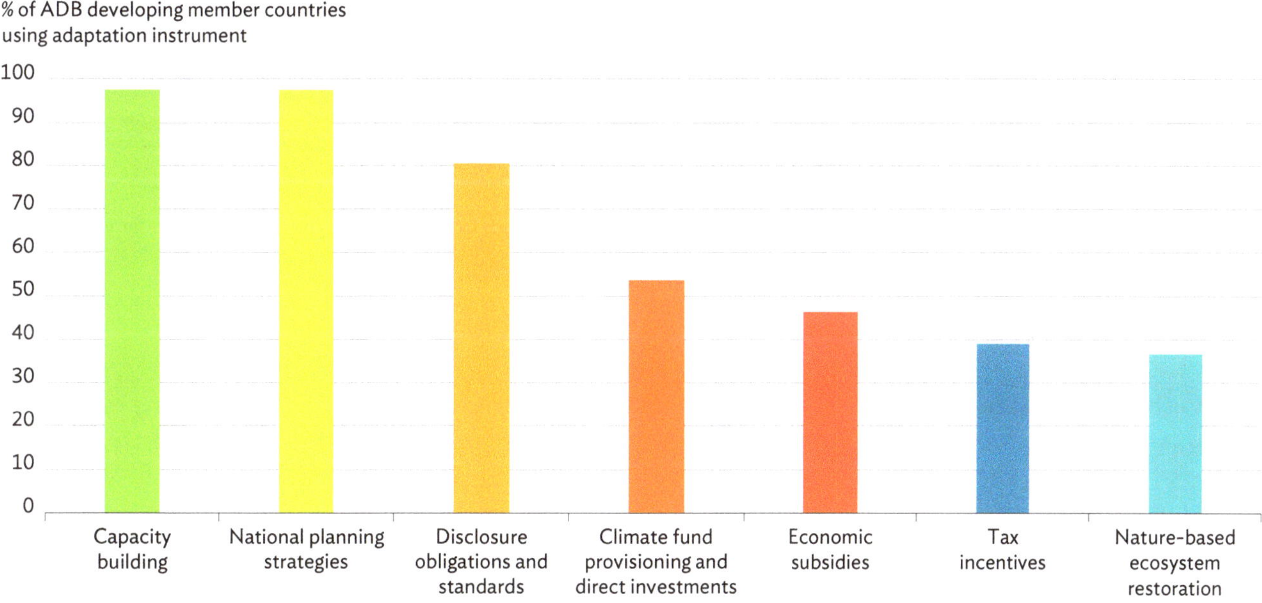

Figure 3.7: Key Adaptation Policy Instruments Across Asia and the Pacific

Notes: The bar chart represents an analysis of policy instruments employed by the Asian Development Bank's developing member economies to support adaptation, based on the Climate-Laws database as of May 2024. Data cover 41 Asian Development Bank developing member economies, excluding Brunei Darussalam; Hong Kong, China; the Republic of Korea; Singapore; and Taipei,China.
Source: Markandya, A., P. Orgen, and S. Abdulla. 2024. Adaptation Responses to Date in Asia-Pacific Region. Background paper for the *Asia–Pacific Climate Report 2024: Catalyzing Finance and Policy Solutions*. Asian Development Bank.

Box 3.6: Digital Technologies to Contribute to Climate Adaptation

Digital technologies can be powerful tools that can enhance climate adaptation efforts across the Asia and Pacific region. In order to effectively adapt to climate change and allow policymakers to make timely and informed decisions, countries of the region must be able to identify, measure, and manage the risks arising from climate impacts. This process can be supported by digital measurement, reporting, and verification systems, which can be highly effective at defining, quantifying, and monitoring climate risks. To this end, a number of breakthrough and enabling technologies have the potential to amplify adaptation and resilience efforts across Asia and the Pacific.

Countries may leverage artificial intelligence to improve weather forecasting. For example, ocean surface temperatures can now be measured and integrated into models to enhance the understanding and near-term prediction of ocean speeds and currents.

Meanwhile, geospatial and aerial systems are increasingly being used to detect anomalies and support early warning information for climate-related emergencies. India, for example, has deployed multihazard early warning and "dewarning" systems. Sensor technologies can be deployed to collect data on wind speed, air pressure, rainfall, and water levels, and to communicate with Internet of Things technologies to report on potential storm surges and/or rapidly rising water levels. Such technologies can help predict potential flooding in time for remedial action to be taken.

Digital technologies also can help governments improve the efficiency and effectiveness of existing policies and programs. For example, the use of satellite imagery, some of which is freely available to all economies in Asia and the Pacific, can help substantially reduce the costs involved in monitoring and scheduling irrigation. Digital payment systems can enable access to funds in times of shocks and can facilitate adaptive social protection programs and relief efforts. Smart metering for power and water connections can enable pricing based on scarcity that encourages efficient use when demand is high relative to supply during heatwaves or droughts. It can also enable feedback to consumers on how to improve efficiency.

Sources: Orgen, P. and S. Abdulla. 2024. Building Resilience: Adaptation Investment Needs in the Asia–Pacific Region. Background paper for the *Asia–Pacific Climate Report 2024: Catalyzing Finance and Policy Solutions.* Asian Development Bank; and Technology Needs Assessment. TNA Database (accessed 1 May 2024).

3.3.2 Slow Progress in Meeting Adaptation Financing Needs

Progress has been made on scaling up financial resources for climate adaptation, but financing remains inadequate. In 2009, at the 15th Conference of the Parties to the UNFCCC (COP15 in Copenhagen), developed economies committed to a collective goal of mobilizing $100 billion in public and private finance per year by 2020 for climate action (mitigation and adaptation) in developing countries. Progress toward this goal has been tracked by the OECD. While the goal was not met by 2020, it was achieved in 2022 when developed countries provided and mobilized a total of $116 billion in climate finance for developing countries (OECD 2024). The majority (80%) of this finance was public climate finance (bilateral and multilateral) and most of it (65%) was mobilized for mitigation. Global international adaptation finance was $32 billion in 2022. In November 2021, at COP26 in Glasgow, a decision was taken to urge developed countries to at least double (from 2019 levels) their collective provision of adaptation finance to developing countries by 2025. This would increase adaptation finance flows to developing countries to approximately $40 billion per year. At COP28 in Dubai, agreement was also reached to create a new loss and damage fund to assist countries impacted by climate change.

The new collective goal for climate finance is crucial for vulnerable countries in Asia and the Pacific. At COP21 in Paris, it was decided to establish, before the end of 2024, a New Collective Quantified Goal for climate finance. This new target has a floor of $100 billion per year and must account for the needs and priorities of developing economies. With formal deliberations ongoing as of September 2024, the new goal will be especially important for the Asia and Pacific region—particularly the small island developing states, low-income economies, and least-developed countries in the region, due to their high vulnerability and higher relative needs for climate adaptation financing.

Investment needs have been generated by sectoral models and expressed by many countries.
Quantifying the amount of finance needed for adaptation in developing countries is challenging. One method, as used in the Adaptation Gap Report (UNEP 2023) is to use sectoral models. These models first assess the potential impacts of climate change and then specify the adaptation measures (and their associated costs) needed to reduce those impacts to acceptable levels. An alternative approach is to use the adaptation priorities and their associated financing needs as submitted in NDCs and NAPs. It should be noted here that, in general terms, use of the first method mostly (but not always) generates lower values relative to the second method. Neither method can be considered superior to the other. The international modeling approach has consistency across countries and a transparent basis, but it is restricted to a limited set of considered options. National needs vary in quantification methods, assumptions and rigor across countries, but may include options and specificities that international modeling misses.

The adaptation finance needed for Asia and the Pacific, based on modeling, is $102 billion per year.
An indicative analysis of climate adaptation finance needed in the Asia and Pacific region, based on sector modeling (UNEP 2023), reveals a central value of $102 billion per year through to 2030 (Figures 3.3 and 3.8). The majority of costs are in the PRC and India. These modeled costs are based on the additional adaptation action needed to address modeled climate impacts. The largest total modeled costs are for the subregion of East Asia, predominantly due to adaptation needs in the PRC. However, in terms of modeled costs per capita, the highest values are for the small island developing states in the Pacific (Figure 3.9). These economies are characterized by geographic isolation and relatively small populations, with high exposure to climate change, low adaptive capacities, and high transportation costs for energy and materials used in adaptation investments.

Figure 3.8: Stated and Modeled Annual Adaptation Finance Needs for 2023–2030, Total

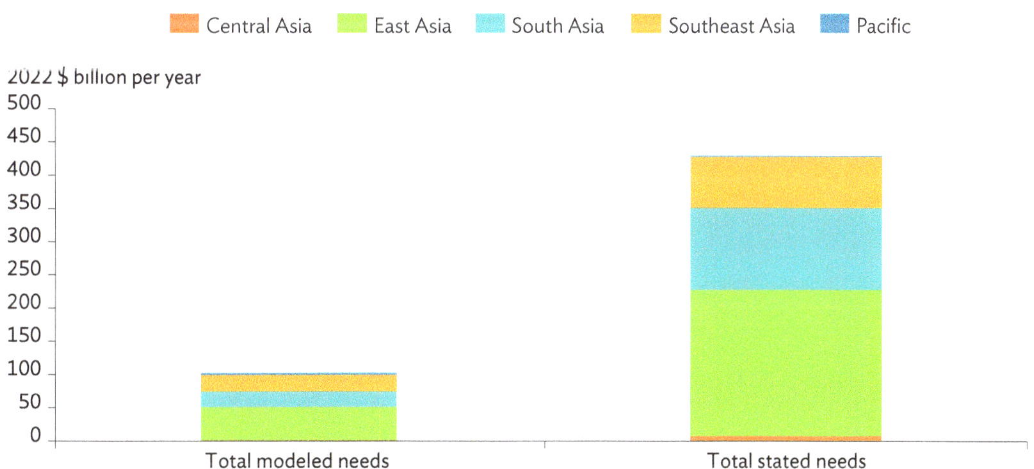

Notes: References for stated climate adaptation needs are, in order of priority, (i) National Adaptation Plans, (ii) nationally determined contributions, (iii) National Communications on Climate Change, (iv) adaptation communications, and (v) average per capita values per income class for economies that have not expressed adaptation needs, following the methodology described in the United Nations Environment Programme's Adaptation Gap Report 2023. Central Asia comprises Armenia, Azerbaijan, Georgia, Kazakhstan, the Kyrgyz Republic, Tajikistan, Turkmenistan, and Uzbekistan. East Asia includes the People's Republic of China and Mongolia. The Pacific comprises the Cook Islands, the Federated States of Micronesia, Fiji, Kiribati, the Marshall Islands, Nauru, Niue, Palau, Papua New Guinea, Samoa, Solomon Islands, Tonga, Tuvalu, and Vanuatu. South Asia comprises Afghanistan, Bangladesh, Bhutan, India, Maldives, Nepal, Pakistan, and Sri Lanka. Southeast Asia comprises Cambodia, Indonesia, the Lao People's Democratic Republic, Malaysia, Myanmar, the Philippines, Thailand, Timor-Leste, and Viet Nam.
Sources: Modeled needs are authors' calculations from data assembled for United Nations Environment Programme. 2023. *Adaptation Gap Report 2023: Underfinanced. Underprepared. Inadequate Investment and Planning on Climate Adaptation Leaves World Exposed.* Stated needs are from Markandya, A., P. Orgen, and S. Abdulla. 2024. Adaptation Responses to Date in the Asia-Pacific Region. Background paper for the *Asia–Pacific Climate Report 2024: Catalyzing Finance and Policy Solutions.* Asian Development Bank.

An updated analysis of stated adaptation finance needs indicates a requirement for $431 billion per year to 2030. Based on submitted NDCs, NAPs, adaptation communications, and NCCCs, stated adaptation finance needs provide valuable insights and are an important source of evidence for estimating global adaptation finance needs and, in turn, calculating the adaptation finance gap. East Asia reports the highest subregional needs, predominantly due to the PRC. Beyond this, on a per capita level, many needs are of a similar magnitude across subregions (Figure 3.9). However, as above, is important to note that the cost information provided in these national reports is inconsistent with respect to scope and scenario, and that the methods employed to prioritize and cost adaptation options vary (UNFCCC Adaptation Committee 2022). Stated needs may be higher than modeled needs for various reasons, including assumption of greater future warming than in the modeled scenarios, inclusion of sectors that were not modeled or broader consideration of needs in the sectors, incorporation of blended development and climate adaptation needs, or differences in methods.

Figure 3.9: Stated and Modeled Annual Adaptation Finance Needs for 2023–2030, per Capita

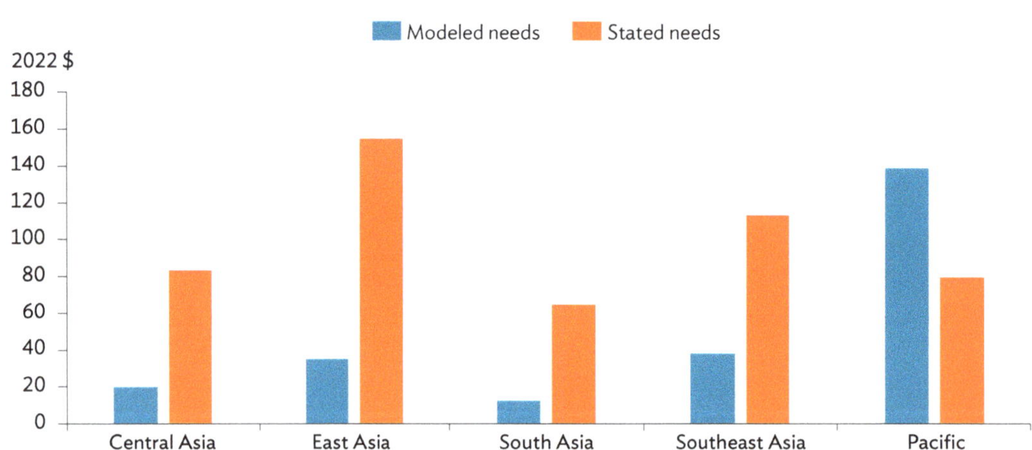

Notes: References for stated climate adaptation needs are, in order of priority, (i) National Adaptation Plans, (ii) nationally determined contributions, (iii) National Communications on Climate Change, (iv) adaptation communications, and (v) average per capita values per income class for countries that have not expressed adaptation needs, following the methodology described in the United Nations Environment Programme's Adaptation Gap Report 2023. Central Asia comprises Armenia, Azerbaijan, Georgia, Kazakhstan, the Kyrgyz Republic, Tajikistan, Turkmenistan, and Uzbekistan. East Asia includes the People's Republic of China and Mongolia. The Pacific comprises the Cook Islands, the Federated States of Micronesia, Fiji, Kiribati, the Marshall Islands, Nauru, Niue, Palau, Papua New Guinea, Samoa, Solomon Islands, Tonga, Tuvalu, and Vanuatu. South Asia comprises Afghanistan, Bangladesh, Bhutan, India, Maldives, Nepal, Pakistan, and Sri Lanka. Southeast Asia comprises Cambodia, Indonesia, the Lao People's Democratic Republic, Malaysia, Myanmar, the Philippines, Thailand, Timor-Leste, and Viet Nam.
Sources: Modeled needs are from United Nations Environment Programme. 2023. *Adaptation Gap Report 2023: Underfinanced. Underprepared. Inadequate Investment and Planning on Climate Adaptation Leaves World Exposed.* Stated needs are from Markandya, A., P. Orgen, and S. Abdulla. 2024. Adaptation Responses to Date in the Asia-Pacific Region. Background paper for the *Asia–Pacific Climate Report 2024: Catalyzing Finance and Policy Solutions.* Asian Development Bank.

Regardless of method for estimating needs, the gap between adaptation needs and finance flows is enormous for the Asia and Pacific region. Total tracked adaptation commitments for the region amounted to $34 billion annually in 2021–2022, of which $24 billion was domestic expenditure that was primarily in the PRC (GCA and CPI 2024). This compares to estimated needs in the range of $102 billion–$431 billion per year. It is likely that there are additional untracked financial commitments to climate adaptation. Nonetheless, even if they could be included, the adaptation finance gap for Asia and the Pacific would likely remain in the hundreds of billions of dollars annually, based on financing needs stated in the region's UNFCCC submissions, or many tens of billions based on modeled needs. International adaptation finance flows to Asia and the Pacific were approximately $11 billion in 2021–2022.

3.3.2.1 Options to Close the Finance Gap

International finance can be accelerated, but it will not solve the financing gap alone. Global international climate adaptation finance has grown threefold from 2016 to 2022, as international financial institutions and donor countries have increasingly targeted providing adaptation finance. While international adaptation finance can continue to grow rapidly, feasible levels of growth from $11 billion of finance in 2021–2022 will not soon attain a large share of the financing needs of $102 billion or more annually for the Asia and Pacific region. Moreover, existing financing flows tend to be more conditioned by country capacity to prepare and implement projects for finance than the climate vulnerability of the recipient country (Jones et al. 2024). To help this financing flow to the countries most in need, special assistance is needed to help vulnerable countries to build adaptation project planning and implementation processes, including disbursement capacity.

Private investment accounts for less than 1% of adaptation finance flows to developing Asia and the Pacific, as adaptation is in some ways harder to finance than mitigation. The low level of adaptation finance flows relative to mitigation (e.g., 31% versus 65% of international public climate finance flows in 2022) partly reflects the fact that climate adaptation measures face special financing challenges. While, intuitively, mitigation may seem more challenging, as it produces global public goods rather than the local benefits of adaptation, the actions involved are more amenable to financing. Mitigation is primarily technical in nature, often involving large investments in renewable energy that can be made attractive through regulations, subsidies, blended finance, and other fiscal policies, with returns that are only subject to more technological and policy uncertainties than conventional investments. Adaptation involves many more uncertainties around global climate policy, climate science, and socioeconomic development. As it is intertwined with development, climate adaptation is more difficult to label, track, and package to investors, and the benefits generated are often not "green" either. Furthermore, adaptation often has "public good" characteristics and is typically delivered by the public sector (e.g., flood protection) or is in nonmarket sectors. It often requires much greater focus on process and soft measures and, where technology is involved, it is more site- and context-specific, so there are challenges in replicability and capacity to scale up. Perhaps the greatest challenge is that many adaptation measures do not create revenue streams or deliver cost savings, making it difficult to repay finance or attract private investors. As a result, most adaptation finance is funded or financed by the public sector, domestically or internationally. Indeed, Orgen and Abdulla (2024) identify that private adaptation finance accounted for less than 1% of extended finance in the Asia and Pacific region between 2011 and 2021.

To date, adaptation has mostly been financed using grants and loans, but there are opportunities for innovative financial instruments to attract private capital. The pool of financial instruments can be expanded to "de-risk" private sector climate adaptation investments. Governments and international financial institutions are strategically positioned to employ catalytic instruments, such as blending private and/or public finance (e.g., resilience bonds, debt-for-nature swaps) to fund adaptation priorities and reduce upfront capital and life-cycle operations and lower maintenance costs. Catalytic investments can provide capital that is patient, risk-tolerant, concessionary, and flexible, differing from conventional investments by de-risking and mobilizing additional investment from mainstream investors. Risk guarantees can cover repayment obligations under certain conditions or provide the economic incentives needed to attract private capital. These guarantees may support hybrid ordinary-concessional funding frameworks and develop bankable products for the private sector. Financial disclosure requirements regarding climate risks can also create incentives to ensure that private investment does not exacerbate risks and lead to climate resilience.

Public–private partnerships offer another mode for adaptation provision. These financing arrangements enable the private sector to support the provision of public goods. Public–private partnerships leverage the strengths of both the public sector (e.g., accountability, stability) and the private sector (e.g., innovation, efficiency, competition) to provide a balanced and more attractive investment environment (House 2016). While the public sector's participation may help lower initial and life-cycle investment costs (e.g., through minimum revenue guarantees and managerial flexibility), the private sector typically assumes the risks associated with construction, operation, and financing under long-term contracts.

Domestic finance will be important. While international negotiations place the onus on developed countries to provide adaptation funding, governments and development partners will likely need to mobilize domestic public and private resources to adequately and sustainably fund climate adaptation goals (UNEP 2023). These moves could include reforming tax expenditures to free fiscal space in country budgets to pursue adaptation. They may also involve removing or reducing inefficient fiscal expenditures, such as fossil fuel subsidies, to align tax expenditures with adaptation goals. Governments might also elect to develop national strategies for engaging the private sector, including a greater focus on country investment platforms that focus on adaptation as well as revenue recycling of carbon market revenues to support adaptation finance. It is clear that many developing countries are already financing adaptation (UNFCCC 2022), as identified in Climate Budget Tagging and Climate Public Expenditure and Institutional Review studies, though the estimates in these studies are indicative. Further improvement of these processes and estimates could be an incentive to develop new revenue-raising approaches for adaptation.

Innovative revenue-generating mechanisms could help to finance climate adaptation. Innovative revenue-generating mechanisms could include increasing adaptation finance quotas from carbon pricing mechanisms and establishing premium credits specifically for voluntary carbon projects that incorporate climate adaptation and resilience components. Such mechanisms can help attract private sector investment and provide more sustainable funding sources for adaptation projects.

Adaptation investment planning can help develop bankable pipelines of adaptation investments. For many developing countries, implementation strategies for their NAPs lack sufficient detail on financial mobilization arrangements, implementation costs, and how plans will be implemented (Orgen and Abdulla 2024). As a result, there is a need to move beyond normative plans and develop more bankable pipelines of climate adaptation. One way of addressing this is through adaptation investment planning—which seeks to fill the informational gap between NAPs and the detail needed for bankable adaptation investments—at the strategic rather than project level (ADB 2023c). Effective adaptation investment planning involves a greater degree of economic analysis and financial assessment, with more focus on how to develop public, private, and blended finance solutions, as well as considering the needs of different market participants in financing adaptation (i.e., what the private sector can do and enabling conditions needed).

3.4   Policy Recommendations to Enhance Adaptation Outcomes

There is no avoiding the impacts of climate change, so stronger policy responses are needed to minimize loss and damage. As shown in Chapter 2 of this report, climate change poses a substantial risk to development in the Asia and Pacific region, which has not yet met commensurate responses. While countries have made progress in identifying adaptation priorities and needs, relevant implementation has moved more slowly. Actions are needed in a wide range of sectors, ranging from water, to agriculture, energy, social protection, education, and health. All of these actions depend on better assessing and communicating climate risk, more systematic planning of adaptation investments, and mobilizing additional finance.

3.4.1   Better Assessing and Communicating Climate Risk

Appropriate adaptation requires continuous reassessment of potential climate change impacts. More proactive and anticipatory government approaches to climate adaptation must be fluid because of the high uncertainty around future climate change. Climate models are continually improving and the understanding of impacts is evolving, so there is a need for policymakers to continually revisit and improve predictions with new science, including via updated climate models, downscaling models, sector models, and economic models. This requires strengthened capacity to develop and interpret a range of projections.

Decision processes can regularly reflect risk appraisal. A standardized approach to assess, communicate, and disclose climate risk information is critical for guiding decision-making across all sectors and levels of government. Capacity of national and local institutions can be strengthened for data collection to improve the accuracy of localized climate projections, information on exposure of assets, and population and information on localized factors contributing to socioeconomic vulnerabilities. In turn, this can enable improved multihazard risk assessments.

Climate risk communication can be improved. More decisions can be informed by assessing, communicating, and disclosing climate risk information in a format that can be used by different sector agencies, subnational governments, private sector entities, and local communities. As discussed in Chapter 4 of this report, climate risk disclosures can be required as part of wider financial disclosure processes to incentivize the private sector to internalize risks and align financial flows with adaptation needs.

3.4.2   Planning, Prioritizing, and Appraising Adaptation Investments More Systematically

Adaptation can be integrated into national development and expenditure planning to help catalyze climate-resilient development. Such integration can include medium-term plans, investment strategies, and medium-term expenditure frameworks (OECD 2015). This requires a more programmatic and strategic approach to adaptation, rather than implementation as stand-alone projects. These national and sector planning processes provide a key entry point for climate adaptation at scale, linking adaptation to development objectives and key performance indicators in line ministries, leveraging development budgets, and shifting national programming along more climate-resilient pathways. In turn, this requires the involvement of relevant finance, economics, and planning ministries (Watkiss and Cimato 2016; WRI 2018).

Adaptation actions can be mainstreamed across policy and operational levels and aligned with key performance indicators of different sectors. To ensure adaptation outcomes, climate risk and adaptation considerations should be appropriately reflected in national and local development plans and sectoral strategies. This can be accomplished by establishing clear, science-based criteria to prioritize and sequence adaptation projects, focusing on assessment of climate risks (section 3.4.1), cost-effectiveness, and co-benefits. This can involve utilizing multi-criteria decision analysis and other economic tools to evaluate climate adaptation options and ensuring that both immediate and long-term impacts are assessed. In addition, adaptation considerations can be mainstreamed into economic analysis requirements applied by economic planning ministries to discern if additional measures are needed for a range of investments.

Markets can be made to work better for climate adaptation. Certain policies, such as subsidized water prices, obscure climate signals while others, such as distortions to land markets, impede autonomous adaptation responses. Policies can be systematically assessed and reformed to ensure that climate risk is more effectively transmitted into market prices and that individuals are not impeded from appropriately responding to changes in risk.

Institutional capacity can be strengthened to enable effective adaptation policy. Given that climate change affects all sectors, effective policy often depends on strengthening institutional frameworks to support cross-sectoral coordination and collaboration in adaptation planning and implementation. Adaptation interventions involve substantial uncertainty, both around the climate risks to be addressed, and the effectiveness of solutions. In this context, developing robust monitoring, evaluation, and learning mechanisms to track the effectiveness of climate adaptation investments, adjusting plans dynamically, and promoting accountability, are essential actions to ensure effectiveness.

3.4.3   Securing and Sustaining Expanded Finance for Adaptation

Financing remains a key barrier to enabling effective climate adaptation. The estimated adaptation funds needed in Asia and the Pacific to reduce projected climate impacts are many times higher than the current flows of public finance. Given the scale of needs, even if international public finance rises rapidly, other sources of finance will still be needed, including from domestic public and private sector finance, although the opportunity for these will vary by economy (UNEP 2023). A range of adaptation needs—including infrastructure, capacity building, institutional strengthening, and governance—depend on mobilizing this new investment.

Concessional finance can be enhanced. One solution is to expand programmatic access to climate finance through multilateral climate funds, while building capacity in vulnerable countries to propose projects and streamlining national procedures to ensure efficient fund disbursement. However, these measures will still only address a tiny share of needs. Given its scarcity, the limited concessional finance available needs to deliver maximum impact, whether this lies in producing the greatest catalytic effect for key public investment (especially in climate adaptation measures that are difficult to finance) or in leveraging private sector flows.

Financing can be diversified. More broadly, blended finance models can combine public, private, and philanthropic capital to increase the scale and reach of climate adaptation investments. New financial instruments, such as resilience bonds and disaster risk insurance, can reduce risk and manage residual risk, thereby protecting assets and vulnerable communities. Domestic resource mobilization can be strengthened, as appropriate adaptation has high cost–benefit ratios. Different funding sources for adaptation, including taxation and fees, can be utilized to minimize fiscal impacts.

References

ADB (Asian Development Bank). 2024. *Asian Economic Integration Report 2024 Decarbonizing Global Value Chains*.

_____. 2023a. Technical Assistance for Increasing Investments in Early Warning Systems to Strengthen Climate and Disaster Resilience.

_____. 2023b. *Asia in the Global Transition to Net Zero: Asian Development Outlook 2023 Thematic Report*.

_____. 2023c. *Climate Adaptation Investment Planning: A Program to Bridge the Gap Between Climate Adaptation Planning and Financing*. Brochure.

Bachner, G., B. Bednar-Friedl, and N. Knittel. 2019. How Does Climate Change Adaptation Affect Public Budgets? Development of an Assessment Framework and a Demonstration for Austria. *Mitigation and Adaptation Strategies for Global Change*. 24. pp. 1325–1341.

Babendreier, D., M. Wan, R. Tang, R. Gu, J. Tambo, Z. Liu, M. Grossrieder, et al. 2019. Impact Assessment of Biological Control-Based Integrated Pest Management in Rice and Maize in the Greater Mekong Subregion. *Insects*. 10 (8). p. 226.

Bangalore, M. R., S. Hallegatte, M. Bangalore, L. Bonzanigo, M. Fay, T. Kane, U. Narloch, J. Rozenberg, D. Treguer, and A. Vogt-Schilb. 2016. Shock Waves: Managing the Impacts of Climate Change on Poverty. *Climate Change and Development Series*. World Bank.

Berrang-Ford, L., A. R. Siders, A. Lesnikowski, A. P. Fischer, M. W. Callaghan, N. R. Haddaway, K. L. Mach, et al. 2021. A Systematic Global Stocktake of Evidence on Human Adaptation to Climate Change. *Nature Climate Change*. 11. pp. 989–1000.

Brady, O. J., D. D. Kharisma, N. N. Wilastonegoro, K. M. O'Reilly, E. Hendrickx, L. S. Bastos, L. L. Yakob, and D. S. Shepard. 2020. The Cost-Effectiveness of Controlling Dengue in Indonesia using wMel Wolbachia Released at Scale: A Modelling Study. *BMC Medicine*. 18. 186.

Brullo, T., J. Barnett, E. Waters, and S. Boulter. 2024. The Enablers of Adaptation: A Systematic Review. *npj Climate Action*. 3. 40.

Burton, I. 2009. *Climate Change and the Adaptation Deficit*. In E. L. F. Schipper and I. Burton, eds. *The Earthscan Reader on Adaptation to Climate Change*. Earthscan.

CCC (Climate Change Committee). 2021. *Independent Assessment of UK Climate Risk*.

CCRIF (Caribbean Catastrophe Risk Insurance Facility). 2010. *Caribbean Catastrophe Risk Insurance Facility: Enhancing the Climate Risk and Adaptation Fact Base for the Caribbean*. Informational brochure.

Cevik, S. and J. T. Jalles. 2020. *This Changes Everything: Climate Shocks and Sovereign Bonds*. International Monetary Fund (IMF).

Chausson, A., B. Turner, D. Seddon, N. Chabaneix, C. A. J. Girardin, V. Kapos, I. Key, et al. 2020. Mapping the Effectiveness of Nature-Based Solutions for Climate Change Adaptation. *Global Change Biology*. 26 (11). pp. 6134–6155.

Colelli, F. P., I. S. Wing, and E. D. Cian. 2023. Air-Conditioning Adoption and Electricity Demand Highlight Climate Change Mitigation–Adaptation Tradeoffs. *Scientific Reports*. 13. 4413.

CRC (Climate Ready Clyde). 2020. *What Does Transformational Adaptation Look Like? Literature Review Synthesis Paper*. P. Watkiss and F. Cimato, eds. Clyde Rebuilt.

DCP (Disease Control Priorities). n.d.

DEFRA (Government of the United Kingdom, Department of Environment Food and Rural Affairs). 2024. *Accounting for the Effects of Climate Change: Supplementary Green Book Guidance.*

Dercon, S. 2004. Climate Change, Green Growth, and Aid Allocation to Poor Countries. *Oxford Review of Economic Policy*. 30 (3). pp. 436–462.

DFID (Government of the United Kingdom, Department for International Development). 2015. Ethiopia Productive Safety Net Programme Phase 4 (PSNP 4).

Dixon, A. M., P. M. Forster, S. F. Heron, A. M. K. Stoner, and M. Bege. 2022. Future Loss of Local-Scale Thermal Refugia in Coral Reef Ecosystems. *PLOS Climate*. 1 (2).

Duchenne-Moutien, R. A. and H. Neetoo. 2021. Climate Change and Emerging Food Safety Issues: A Review. *Journal of Food Protection*. 84 (11). pp. 1884–1897.

Ebi, K. L., T. J. Teisberg, L. S. Kalkstein, L. Robinson, and R. F. Weiher. 2004. Heat Watch/Warning Systems Save Lives: Estimated Costs and Benefits for Philadelphia 1995–98. *Bulletin of the American Meteorological Society*. 85 (8). pp. 1067–1074.

EIB (European Investment Bank). 2022. *Joint Methodology for Tracking Climate Change Adaptation Finance*.

G20 HLIP (G20 High Level Independent Panel). 2020. *A Global Deal for Our Pandemic Age: Report of the G20 High Level Independent Panel on Financing the Global Commons for Pandemic Preparedness and Response*.

Garbero, A., P. Marion, and V. Brailovskaya. 2022. *The Impact of the Adoption of CGIAR's Improved Varieties on Poverty and Welfare Outcomes*. International Fund for Agricultural Development.

GCA (Global Center on Adaptation) and CPI (Climate Policy Initiative). 2024. *State and Trends in Climate Adaptation Finance 2024*.

GCA (Global Commission on Adaptation). 2019. *Adapt Now: A Global Call For Leadership on Climate Resilience*.

_____. 2021. Macroeconomics and Climate Adaptation. In *State and Trends in Adaptation Report 2021: Africa*.

Giardino, A. and M. Vousdoukas. 2024. Rising Seas: Building Resilience Against Coastal Flooding in Asia and the Pacific. *Development Asia*. 7 August.

Haasnoot, M., J. H. Kwakkel, W. E. Walker, and J. Ter Maat. 2013. Dynamic Adaptive Policy Pathways: A Method for Crafting Robust Decisions for a Deeply Uncertain World. *Journal of Global Environmental Change*. 23 (2). pp. 485–498.

Hallegatte, S., J. Rentschler, and J. Rozenberg. 2019. Lifelines: The Resilient Infrastructure Opportunity. *Sustainable Infrastructure Series*. World Bank.

Hallegatte, S., J. Rozenberg, J. Rentschler, C. Nicolas, and C. Fox. 2019. Strengthening New Infrastructure Assets: A Cost-Benefit Analysis. *Policy Research Working Paper No. 8896*. World Bank.

Horton, S. 2017. Cost-Effectiveness Analysis in Disease Control Priorities (Third Edition). In D. T. Jamison, H. Gelband, S. Horton, P. Jha, R. Laxminarayan, C. N. Mock, and R. Nugent, eds. *Disease Control Priorities: Improving Health and Reducing Poverty*. World Bank.

House, S. 2016. Responsive Regulation for Water PPP: Balancing Commitment and Adaptability in the Face of Uncertainty. *Policy and Society*. 35 (2). pp. 179–191.

Hunt, A., J. Ferguson, M. Baccini, P. Watkiss, and V. Kendrovski. 2016. Climate and Weather Service Provision: Economic Appraisal of Adaptation to Health Impacts. *Climate Services*. 7. pp. 78–86.

IPCC (Intergovernmental Panel on Climate Change). 2014. Climate Change 2014: Impacts, Adaptation, and Vulnerability. Part B: Regional Aspects. Contribution of Working Group II to the Fifth Assessment Report of the Intergovernmental Panel on Climate Change. V. R. Barros, C. B. Field, D. J. Dokken, M. D. Mastrandrea, K. J. Mach, T. E. Bilir, et al., eds. Cambridge University Press.

——. 2022a. Climate Change 2022: Impacts, Adaptation and Vulnerability. Contribution of Working Group II to the Sixth Assessment Report of the IPCC. H.-O. Pörtner, D. C. Roberts, M. Tignor, E. S. Poloczanska, K. Mintenbeck, A. Alegría, et al., eds. Cambridge University Press.

——. 2022b. *The Ocean and Cryosphere in a Changing Climate: Special Report of the Intergovernmental Panel on Climate Change.* H. O. Pörtner, D. C. Roberts, M. Tignor, E. S. Poloczanska, K. Mintenbeck, A. Alegría, et al., eds. Cambridge University Press.

Jones, L. P., J. Banga, B. Notkin, B. Savely, and A. G. J. Brochen. 2024. *Closing the Gap : Trends in Adaptation Finance for Fragile and Conflict-Affected Settings*. World Bank Group.

Kruczkiewicz, A., F. Cian, I. Monasterolo, G. Di Baldassarre, A. Caldas, M. Royz, M. Glasscoe, N. Ranger, and M. van Aalst. 2022. Multiform Flood Risk in a Rapidly Changing World: What We Do Not Do, What We Should, and Why It Matters. *Environmental Research Letters*. 17.

Lenton, T. M., J. Rockström, O. Gaffney, S. Rahmstorf, K. Richardson, and H. J. Schellnhuber. 2019. Climate Tipping Points—Too Risky to Bet Against. *Nature*. 27 November.

Lloyds. 2018. Innovative Finance for Resilient Infrastructure. *Innovation Report 2018: Understanding Risk*.

Mechler, R. 2016. Reviewing Estimates of the Economic Efficiency of Disaster Risk Management: Opportunities and Limitations of Using Risk-Based Cost-Benefit Analysis. *Natural Hazards*. 81. pp. 2121–2147.

MOM (Government of Singapore, Ministry of Manpower). 2024. Heat Stress Measures for Outdoor Work.

Mullen, J., D. Gray, and J. de Meyer. 2015. *Evaluating the Impact of Capacity Building by AICAR*. Paper prepared for the 59th AARES Annual Conference, Rotorua, New Zealand. 10–13 February 2015.

OECD (Organisation for Economic Co-operation and Development). 2015. *Climate Change Risks and Adaptation: Linking Policy and Economics*. OECD Publishing.

——. 2023a. PISA 2022 Results (Volume I): The State of Learning and Equity in Education. 5 December.

_____. 2023b. Policies and Strategies for Climate-Compatible Inclusive Growth and Structural Transformation. Background paper for Discussion.

_____. 2024. Climate Finance Provided and Mobilised by Developed Countries in 2013–2022. OECD Publishing.

Oppenheimer, M., B. C. Glavovic , J. Hinkel, R. van de Wal, A. K. Magnan, A. Abd-Elgawad, R. Cai, et al. 2019. Sea Level Rise and Implications for Low-lying Islands, Coasts, and Communities. In H. O. Pörtner, D. C. Roberts, et al., eds. *IPCC Special Report on the Ocean and Cryosphere in a Changing Climate.* Cambridge University Press.

Orgen, P. and S. Abdulla. 2024. Building Resilience: Adaptation Investment Needs in the Asia-Pacific Region. Background paper for the *Asia–Pacific Climate Report 2024: Catalyzing Finance and Policy Solutions.* ADB.

Parrado, R., F. Bosello, E. Delpiazzo, J. Hinkel, D. Lincke, and S. Brown. 2020. Fiscal Effects and the Potential Implications on Economic Growth of Sea-level Rise Impacts and Coastal Zone Protection. *Climatic Change.* 160. pp. 283–302.

PCMSC (Pacific Coastal and Marine Science Center). 2022. Role of Reefs in Coastal Protection. USGS. 27 June.

Phadera, L., H. Michelson, A. Winter-Nelson, and P. Goldsmith. 2019. Do Asset Transfers Build Household Resilience? *Journal of Development Economics.* 138. pp. 205–227.

Raghupathi, V. and W. Raghupathi. 2020. The Influence of Education on Health: An Empirical Assessment of OECD Countries for the Period 1995–2015. *Archives of Public Health.* 78 (20).

Ralaidovy, A. H., J. A. Lauer, C. Pretorius, O. J. T. Briët, and E. Patouillard. 2021. Priority Setting in HIV, Tuberculosis, and Malaria – New Cost-Effectiveness Results from WHO-CHOICE. *International Journal of Health Policy and Management.* 10. pp. 678–696.

Ranger, N., O. Mahul, and I. Monasterolo. 2022. *Assessing Financial Risks from Physical Climate Shocks: A Framework for Scenario Generation.* World Bank.

Rosegrant, M. W., B. Wong, T. B. Sulser, N. Dubosse, and T. J. Lybbert. 2023. Benefit-Cost Analysis of Increased Funding for Agricultural Research and Development in the Global South. *Journal of Benefit-Cost Analysis.* 14 (S1). pp. 1–25.

Rouillard, J., J. Tröltzsch, J. Tarpey, M. Lago, P. Watkiss, A. Hunt, F. Bosello, et al. 2016. The Economic Analysis of Climate Adaptation: Insights for Policymakers. ECONADAPT Deliverable 10.3.

Roy, T. and K. J. George. 2020. Precision Farming: A Step Towards Sustainable Climate-Smart Agriculture. In V. Venkatramanan, S. Shah, and R. Prasad, eds. *Global Climate Change: Resilient and Smart Agriculture.* Springer.

Scheelbeek, P. F. D., A. D. Dangour, S. Jarmul, G. Turner, A. J. Sietsma, J. C. Minx, M. Callaghan, I. Ajibade, S. E. Austin, and R. Biesbroek. 2021. The Effects on Public Health of Climate Change Adaptation Responses: A Systematic Review of Evidence from Low- and Middle-income Countries. *Environmental Research Letters.* 16: 073001.

Seddon, N., A. Chausson, P. Berry, C. A. J. Girardin, A. Smith, and B. Turner. 2020. Understanding the Value and Limits of Nature-Based Solutions to Climate Change and Other Global Challenges. *Philosophical Transactions of the Royal Society B: Biological Sciences.* 375 (1794).

Seneviratne, S. I., X. Zhang, M. Adnan, W. Badi, C. Dereczynski, A. Di Luca, S. Ghosh, et al. 2021. Weather and Climate Extreme Events in a Changing Climate. In *Climate Change 2021: The Physical Science Basis*. Contribution of Working Group I to the Sixth Assessment Report of the IPCC. Cambridge University Press.

Shelton, C. 2014. Climate Change Adaptation in Fisheries and Aquaculture – Compilation of Initial Examples. FAO Fisheries and Aquaculture Circular No. 1088. Food and Agriculture Organization of the United Nations.

Smyth, S. J., S. R. Webb, and P. W. B. Phillips. 2021. The Role of Public-Private Partnerships in Improving Global Food Security. *Global Food Security*. 31. 100588.

Stevenson, J., B. Vanlauwe, K. Macours, N. Johnson, L. Krishnan, F. Place, D. Spielman, K. Hughes, and P. Vlek. 2019. Farmer Adoption of Plot- and Farm-Level Natural Resource Management Practices: Between Rhetoric and Reality. *Global Food Security*. 20. pp. 101–104.

Sulser, T. B., K. D. Wiebe, S. Dunston, N. Cenacchi, A. Nin-Pratt, D. Mason-D'Croz, R. D. Robertson, D. Willenbockel, and M. W. Rosegrant. 2021. *Climate Change and Hunger: Estimating Costs of Adaptation in the Agrifood System*. International Food Policy Research Institute.

Tiggeloven, T., J. Aerts, and P. Ward. 2021. The Benefits of Coastal Adaptation Through Conservation of Foreshore Vegetation. *Journal of Flood Risk Management*.

UNCC (United Nations Climate Change). 2012. Approaches to Address Loss and Damage Associated with Climate Change Impacts in Developing Countries.

————. n.d. National Adaptation Plans.

UNEP (United Nations Environment Programme). 2018. *Climate Change and the Cost of Capital in Developing Countries.*

————. 2023. *Adaptation Gap Report 2023: Underfinanced. Underprepared. Inadequate Investment and Planning on Climate Adaptation Leaves World Exposed.*

UNF (United Nations Foundation). 2023. Submission to the United Nations Framework Convention on Climate Change Glasgow–Sharm el Sheikh Work Programme on the Global Goal on Adaptation from the UN Foundation: Prioritizing Transformational Adaptation.

UNFCCC (United Nations Framework Convention on Climate Change). 2022. *Synthesis Report on the Cost of Adaptation – Efforts of Developing Countries in Assessing and Meeting the Costs of Adaptation: Lessons Learned and Good Practices.*

van der Geest, K. and R. van den Berg. 2021. Slow-Onset Events: A Review of the Evidence from the IPCC Special Reports on Land, Oceans and Cryosphere. *Current Opinion in Environmental Sustainability*. 50.

Venton, C. and L. Coulter. 2013. *The Economics of Early Response and Resilience: Lessons from Niger.*

Volz, U., J. Beirne, N. Ambrosio Preudhomme, A. Fenton, E. Mazzacurati, N. Renzhi, and J. Stampe. 2020. *Climate Change and Sovereign Risk.* SOAS University of London.

Waddington, H., B. Snilstveit, J. Hombrados, M. Vojtkova, D. Phillips, P. Davies, and H. White. 2014. Farmer Field Schools for Improving Farming Practices and Farmer Outcomes: A Systematic Review. International Initiative for Impact Evaluation.

Ward, P., B. Jongman, J. C. J. H. Aerts, P. D. Bates, W. J. W. Botzen, A. Diaz Loaiza, S. Hallegatte, et al. 2017. A Global Framework for Future Costs and Benefits of River-flood Protection in Urban Areas. *Nature Climate Change.* 7. pp. 642–646.

Watkiss, P. and F. Cimato. 2016. *The Economics of Adaptation and Climate-Resilient Development. Lessons from Projects for Key Adaptation Challenges.* Centre for Climate Change Economics and Policy and Grantham Research Institute on Climate Change and the Environment.

Watkiss, P. and R. Betts. 2021. *Climate Change Risk Assessment 3 Evidence Report.* Chapter 2: Method.

Watkiss, P., M. Benzie, and R. J. T. Klein. 2015. The Complementarity and Comparability of Climate Change Adaptation and Mitigation. Wiley Interdisciplinary Reviews. *WIREs Climate Change.* 6. pp. 541–557.

World Bank. 2021. Investment in Disaster Risk Management in Europe Makes Economic Sense. *Background Report: Economics for Disaster Prevention and Preparedness.*

WHO (World Health Organization). 2015. Operational Framework for Building Climate Resilience Health Systems.

WRI (World Resources Institute). 2018. From Planning to Action: Mainstreaming Climate Change Adaptation into Development. Working Paper.

WTO (World Trade Organization). 2022. *World Trade Report 2022: Climate Change and International Trade.*

Ziska, L. H. 2022. Rising Carbon Dioxide and Global Nutrition: Evidence and Action Needed. *Plants.* 11 (7). 1000.

Chapter 4

Scaling Up Private Climate Capital for Mitigation and Adaptation

4.1 Shortfalls in Climate Finance for Asia and the Pacific

Amid urgent efforts to combat climate change, a substantial financing gap still exists. Climate finance has been growing over time, but it has not kept pace with the rapidly rising financing needed to fight climate change. The Climate Policy Initiative has estimated that annual global climate finance must increase from $8.1 trillion in 2023 to $9.0 trillion by 2030 and should exceed $10.0 trillion annually during 2031–2050 (Buchner et al. 2023). In Asia and the Pacific alone, an estimated $2.0 trillion is needed annually during 2022–2030 to meet targets outlined under nationally determined contributions (NDCs) (ADB 2023a).[1] This estimate is well above the average annual amount of $1.3 trillion mobilized globally during 2021–2022 (Buchner et al. 2023).

The private sector needs to play a more prominent role in climate investment. To address this financing gap, Asia and the Pacific can benefit significantly from mobilizing private capital. Estimates show that the private sector accounted for 32% of total climate finance in the region during 2018–2019. However, given the strain on public resources and the competing priorities of governments, the private sector's share in climate finance needs to rise up to 90% by 2030 (IMF 2023).

A more enabling environment is needed to attract private capital and build investor confidence. Transitioning toward low-carbon and climate-resilient development requires significant and sustained changes across the regional economy. In terms of financing, the growing sustainable finance market and the private sector's rising awareness of associated risks and opportunities need to be leveraged. To establish the enabling environment for climate finance to flow, the contexts of individual economies, consistency across current policies, and institutional settings all need to be considered. Climate data availability and market integrity are also key to enhancing investor confidence.

4.2 The Landscape of Private Climate Capital in Asia and the Pacific

Understanding the state of climate finance in Asia and the Pacific is challenging. What constitutes the scope of the region and its economies may vary depending on which geographic or economic criteria are used. The data being recorded are also heavily reliant on reporting approaches that may not be harmonized across economies and institutions and may depend on an economy's resources and capacities. The graphics presented in Figure 4.1 and Figure 4.2 are based on a joint report by the Asian Development Bank (ADB) and the Climate Policy Initiative using cumulative data for 2018 and 2019 (ADB 2023a). Where available, more recent data are presented to supplement the analysis.

---

[1]   It should be noted that estimations vary widely across different models and publications, depending on various assumptions and sample coverage.

This chapter was written by a team consisting of ADB staff Bianca Gutierrez, Esmyra Javier, Ma. Noelle Manahan (consultant), Alexander Raabe, Shu Tian (lead author), and Mai Lin Villaruel.

Globally, the private sector provides almost half of total climate finance. The private sector's contribution to climate finance is driven mostly by mobilization in developed economies. Data from 2021–2022 show that 49% ($625 billion) of climate finance in this period came from the private sector (Buchner et al. 2023). In Asia and the Pacific, data from 2018–2019 show that, of the total $520 billion mobilized for the region,[2] the private sector contributed $168 billion (32%), mainly from corporations, households, and commercial financial institutions (ADB 2023a). Private capital from commercial investors, funds, households, and corporations is channeled through project-level market rate debt, balance sheet debt, and equity. Assessing the subregions of Asia and the Pacific, the private sector's contribution in South Asia was close to the global average at 43% in 2018–2019, while the Pacific's private sector accounted for only 4% (Figure 4.1).

Figure 4.1: Public and Private Sector Shares of Climate Finance Across Asia and the Pacific, 2018–2019

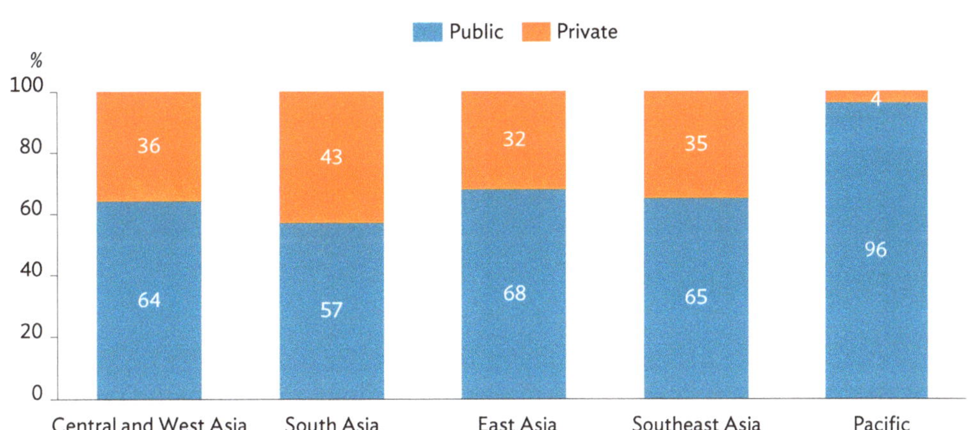

Notes: Percentages are based on cumulative data for 2018 and 2019. When available, more recent data are presented to supplement the analysis. Central and West Asia comprises Armenia, Azerbaijan, Georgia, Kazakhstan, the Kyrgyz Republic, Pakistan, Tajikistan, Turkmenistan, and Uzbekistan. South Asia comprises Bangladesh, Bhutan, India, Maldives, Nepal, and Sri Lanka. East Asia comprises the People's Republic of China and Mongolia. Southeast Asia comprises Cambodia, Indonesia, the Lao People's Democratic Republic, the Philippines, Thailand, Timor-Leste, and Viet Nam. Source: Asian Development Bank. 2023. *Climate Finance Landscape of Asia and the Pacific*.

Most climate finance is mobilized in the form of debt instruments. While grants and equity instruments are also used to fund climate investments, 60% of global climate finance in 2021–2022 was raised through debt instruments. This proportion was slightly higher across Asia and the Pacific, where 70% of climate finance was in the form of debt instruments in 2018–2019. While market rate debt and low-cost project debt instruments were mainly from public institutions, balance sheet debt (which made up 21% of total debt instruments) was raised by corporations. One exception to this trend was in the Pacific, where grants dominated climate finance with a share of 72% (ADB 2023a). By type of activity, energy systems and low-carbon transport received most of the funding across the region (Figure 4.2). Financing sources and instruments for adaptation finance are discussed in greater detail in Chapter 3 of this report.

2　In more recent estimates, total climate finance flowing to Asia and the Pacific region was reported at $746 billion for 2022 (Buchner et al. 2023), but no further breakdown by source was available for the region.

Figure 4.2: Overview of Climate Finance Flows in Asia and the Pacific, 2018–2019
($ billion)

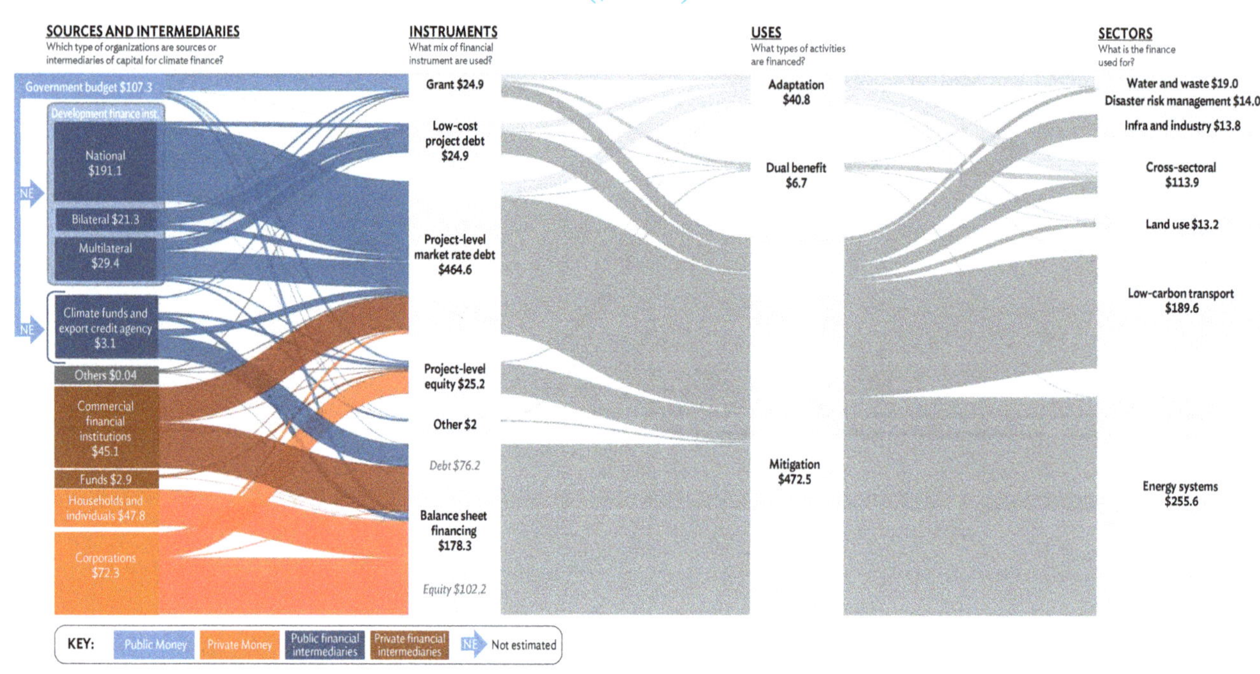

Source: Asian Development Bank. 2023. *Climate Finance Landscape of Asia and the Pacific*.

Consistent with global trends, mitigation finance dominates finance flows in the region. Mitigation finance aims at reducing greenhouse gas (GHG) emissions to slow global warming and stabilize the climate in the long term. Data from 2018–2019 show that 91% of regional climate finance flows were directed to mitigation, with 36% of this being provided by the private sector. Adaptation finance, focused on improving resiliency and reducing the impact of climate-related risk and damage, continues to lag despite the urgency and rising costs of climate change. In Asia and the Pacific, only 8% of total finance flowed to adaptation during 2018–2019 and this was mostly provided by the public sector: the private sector's contribution to adaptation finance was 0.2% or approximately $84 million. Mobilizing private capital for climate change adaptation is often hindered by a lack of bankable projects and scalable business models, making such investments less financially attractive to private investors. Chapter 3 of this report includes detailed discussion on how to mobilize adaptation finance.

Even with higher private sector participation, a funding gap remains for mitigation. Mitigation finance, despite comprising the dominant share of overall climate financing, remains insufficient to meet the targets set forth by economies in Asia and the Pacific in their NDCs. A high level of dependence on fossil fuels— accounting for more than half in global consumption—puts the region in a challenging position in terms of transforming its energy systems and decarbonizing its economies (ADB 2023c). Transition finance is urgently needed in the region, but such financing introduces a different set of complexities in evaluating risks and returns for the private sector (WEF 2023; Kagda 2024; PWC 2023).

Climate investments by the private sector can deliver both financial and nonfinancial gains. An analysis of initial NDC commitments estimated a potential for $23 trillion in investment opportunities from 2016–2030 in emerging markets and across sectors (IFC 2017). Climate policies, development of sustainable financial markets, and de-risking instruments such as guarantees, insurances, and blended finance can be catalytic in attracting private capital. These measures can reduce risks or enhance potential returns and may contribute to nonpecuniary gains such as improved reputation and stakeholder partnerships. Furthermore, when considering the amount of capital deployed as well as capital flows into funds and deal activity, climate-related private investments outperformed the broader market in 2022 (Dahlqvist et al. 2023).

Private sector engagement must be recognized and incorporated in climate policies and implementation plans. Despite the need for increased private sector engagement, it is barely mentioned in public climate commitments as published in documents such as the NDCs, with only 13% of 85 updated NDCs mentioning active private sector engagement in 2021 (Crumpler et al. 2021). Based on submissions to the NDC Registry of the United Nations Framework Convention on Climate Change, 25 of ADB's 41 developing member economies mentioned some form of private sector involvement such as public–private partnerships or creating an enabling environment for the private sector. However, only 14 developing member economies mentioned engaging the private sector, with only one member specifying implementation plans.[3]

Actual and perceived risks of investing in climate finance must be addressed. Regional and economy-level contexts and barriers complicate the climate finance landscape in developing Asia. Barriers such as lack of climate data, resource constraints, less-developed capital markets, and service issues in areas such as electricity and transport constrain private climate investments. Risks relating to the subinvestment-grade ratings of many emerging–market and developing economies, policy and foreign exchange rates, and potential low investment returns also need to be addressed to attract private investment (IMF 2023). Stronger policy commitments and action must be taken to provide de-risking and incentivizing measures. Moreover, an enabling environment that supports private investment considerations must be considered when creating pipelines of policy-backed, investable projects. Likewise, disincentivizing measures, such as more stringent regulatory scrutiny for high-emitting operations or increased taxes for fossil fuel use, can assist in transitioning private firms toward activities aligned with climate goals. Stronger climate commitments, climate policies, and public finance action are crucial to mobilize private climate capital across Asia and the Pacific.

4.3   Drivers of Private Climate Investment

Scaling up private climate capital requires activating both financial and nonfinancial levers. To mobilize much-needed private capital at scale to support mitigation and adaptation, it is important to understand what drives private investment in climate action. These direct and indirect financial drivers can be further categorized into pull factors and push factors (Figure 4.3). Pull factors encourage the private sector to invest in climate finance by leveraging upside opportunities. Push factors tend to impel the private sector to make investments to mitigate downside risks. Over time, push factors may become pull factors as climate investments become an upside opportunity.

---

3   The 14 developing member economies are Bhutan, Cambodia, the People's Republic of China, India, Indonesia, the Lao People's Democratic Republic, Maldives, the Marshall Islands, Sri Lanka, Timor-Leste, Tonga, Turkmenistan, Vanuatu, and Viet Nam. Bhutan specified direct implementation plans through its pursuit of direct access modalities through the accreditation of the Bhutan Trust Fund for Environmental Conservation as the national implementing entity to both the Green Climate Fund and the Adaptation Fund, and private sector access, with three financial institutions (Bhutan Development Bank Limited, Bank of Bhutan Limited, and the Bhutan National Bank Limited) undergoing the accreditation process for access to the private sector facility of the Green Climate Fund (Royal Government of Bhutan 2021).

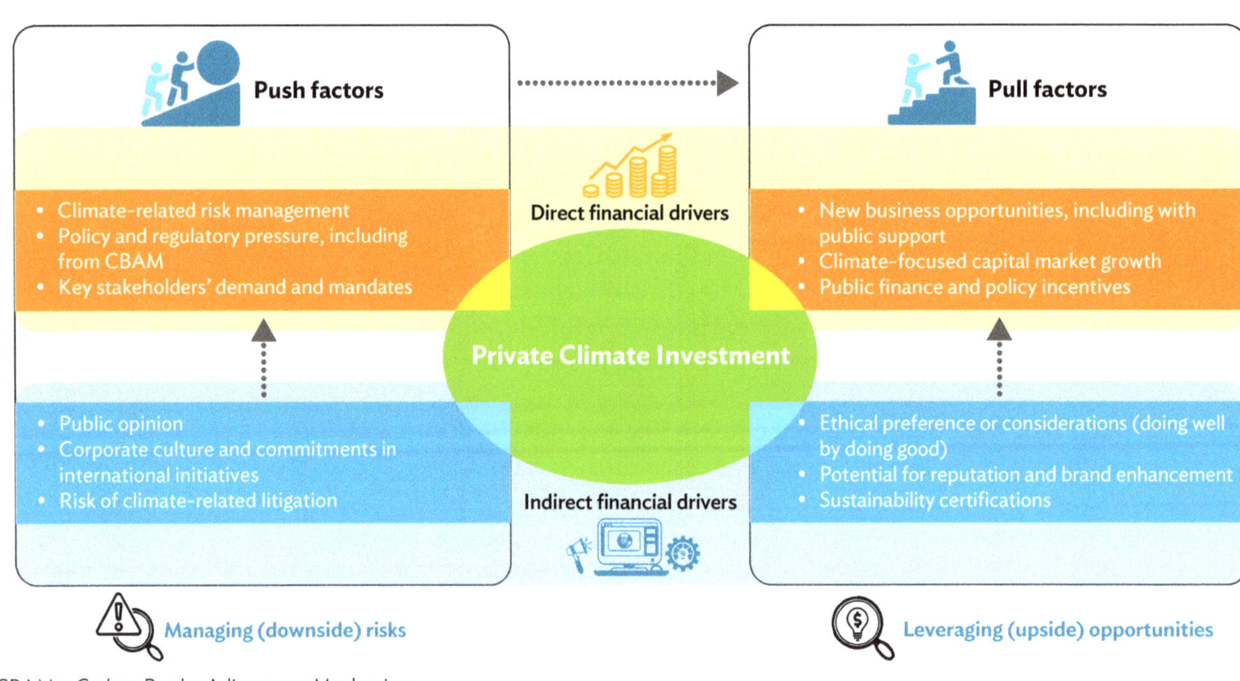

CBAM = Carbon Border Adjustment Mechanism.
Source: Asian Development Bank visualization.

4.3.1   Direct Financial Drivers

Both investment opportunities and financial risks motivate firms to invest in mitigation and adaptation. Key pull factors include profitability, access to capital markets, and climate policy incentives. Some of the key push factors are business continuity, incentivizing climate policies such as carbon pricing, tax and subsidy reform, and shareholder demands.

4.3.1.1   Pull Factors

Tackling climate change has been recognized as an important investment opportunity. Studies by the Organisation for Economic Co-operation and Development (OECD 2023a), Tang (2023), and Wu et al. (2023) conclude that climate finance initiatives offer major investment opportunities. For instance, sustainable investment funds have consistently outperformed their conventional peers, with a median return of 12.6% compared to traditional funds' 8.6% in 2023 (MSISI 2024). This return differential is more pronounced in Asia, with sustainable investment funds providing a 1.4% return while traditional funds reported a 0.4% loss in 2023. Global investment in the low-carbon energy transition hit a record $1.8 trillion in 2023, up by 17% from 2022 (Bloomberg NEF 2024). For opportunities related to mitigation solutions, under the net zero emissions scenario of the International Energy Agency, there are considerable investment opportunities in electricity networks; sustainable vehicles; electric vehicle charging infrastructure; energy generation via wind, solar, hydrogen, and biofuels; raw materials such as copper and lithium; and building heating solutions (Figure 4.4).[4]

---

4   The electrification of transport and other sectors will increase the demand for raw materials such as copper and lithium, whose supply is exposed to regulatory and market risks, presenting opportunities for innovations in materials and recycling solutions (Tang 2023).

Similarly, financing adaptation is increasingly becoming a business opportunity and is estimated to be worth $2 trillion annually by 2030 (The Economist 2021; WEF 2022; Chidambaram and Khanna 2022). For instance, investments in climate-resilient economies and water-saving technologies can generate above average returns (Wu et al. 2023). Businesses providing climate resilience solutions are emerging in Asia and the Pacific (Box 4.1), as also highlighted in Chapter 3 of this report.

Figure 4.4: 2030 Decarbonization Opportunities Under the International Energy Agency's Net Zero Emissions Scenario

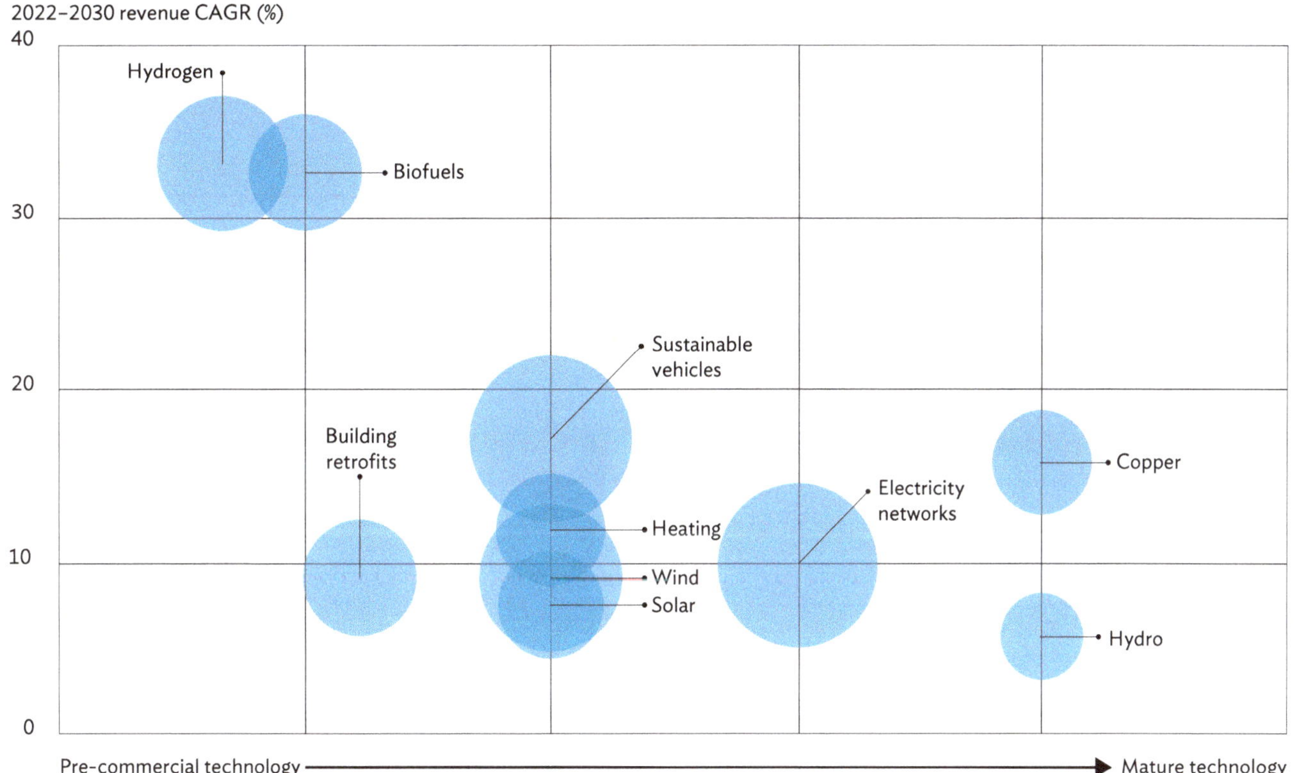

CAGR = compound annual growth rate.
Notes: Bubbles represent CAGRs of sectors with expected revenues over $400 billion by 2030. Indicative representation, not drawn to scale.
Source: Adapted from Figure 2 of Tang, V. 2023. Beyond Financing Gaps: Sizing the Decarbonization Investment Opportunity. Government of Singapore Investment Corporation.

Box 4.1: Climate Resilience Solutions in Asia and the Pacific

Incorporated in India in 2016, Absolute Water Private Limited developed the 100% organic water recovery system known as Absolute Water. The system treats and converts sewage wastewater into potable water, in line with World Health Organization standards, by naturally degrading pollutants and converting them into nutrients. As shown in the figure below, through its applications in industrial processes, residential activities, construction sites, and agricultural land use, the Absolute Water system increases the availability and quality of water in localities subject to water stress.

The Absolute Water Wastewater Treatment System

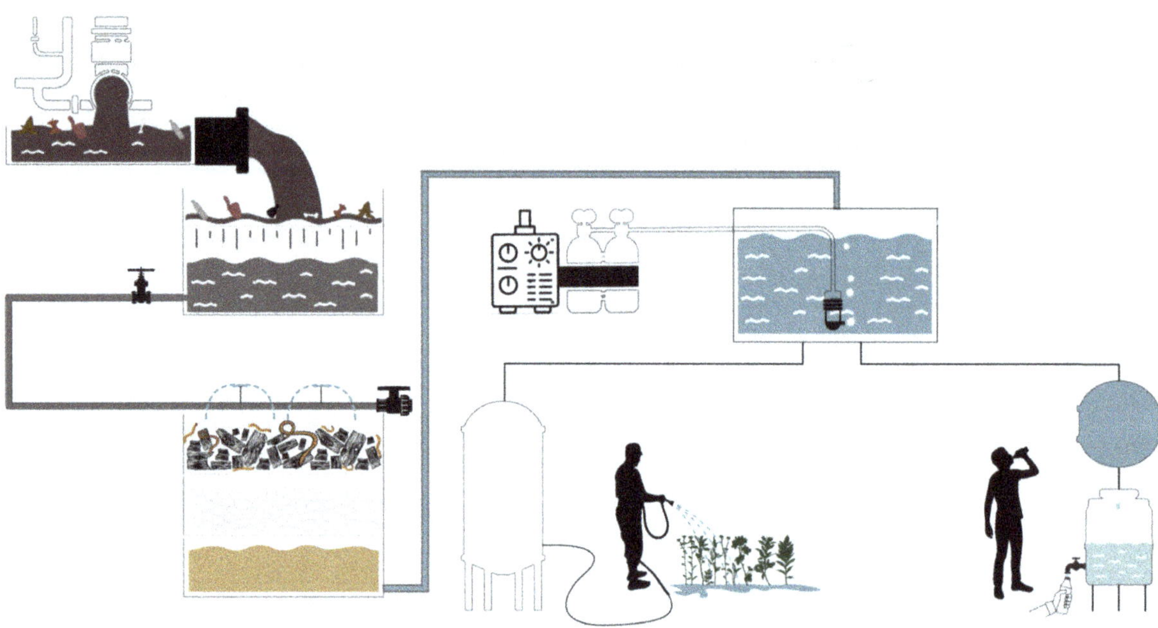

Source: Absolute Water.

Equity and debt capital markets are important conduits to raise climate finance. Private equity increasingly turns into a significant lever to finance the "just transition" to an inclusive, low-carbon, and climate-resilient development pathway (Robins, Brunsting, and Wood 2018).[5] The climate-related private equity market saw sustained growth from 2019 to 2022, buoyed by the increased deployment of climate technologies (Dahlqvist et al. 2023). During the period, transactions increased 2.5 times (twice the growth rate of all private equity market transactions) and investors launched about 330 new funds covering sustainability; environmental, social, and governance (ESG) issues; and climate impacts. Companies operating in renewable energy, clean technology, and "green" infrastructure are particularly attractive investment targets (Calster, Vandenberghe, and Reins 2022), leading to a concentration of climate-related private equity investments in the energy and transport sectors (Figure 4.5).

---

[5]   The International Labour Organization defines "just transition" as "a process by which economies that progress toward a green economy also strengthen each of the four pillars of decent work for all (i.e., social dialogue, social protection, rights at work, and employment)."

Figure 4.5: Climate-Related Private Equity Investments by Sector

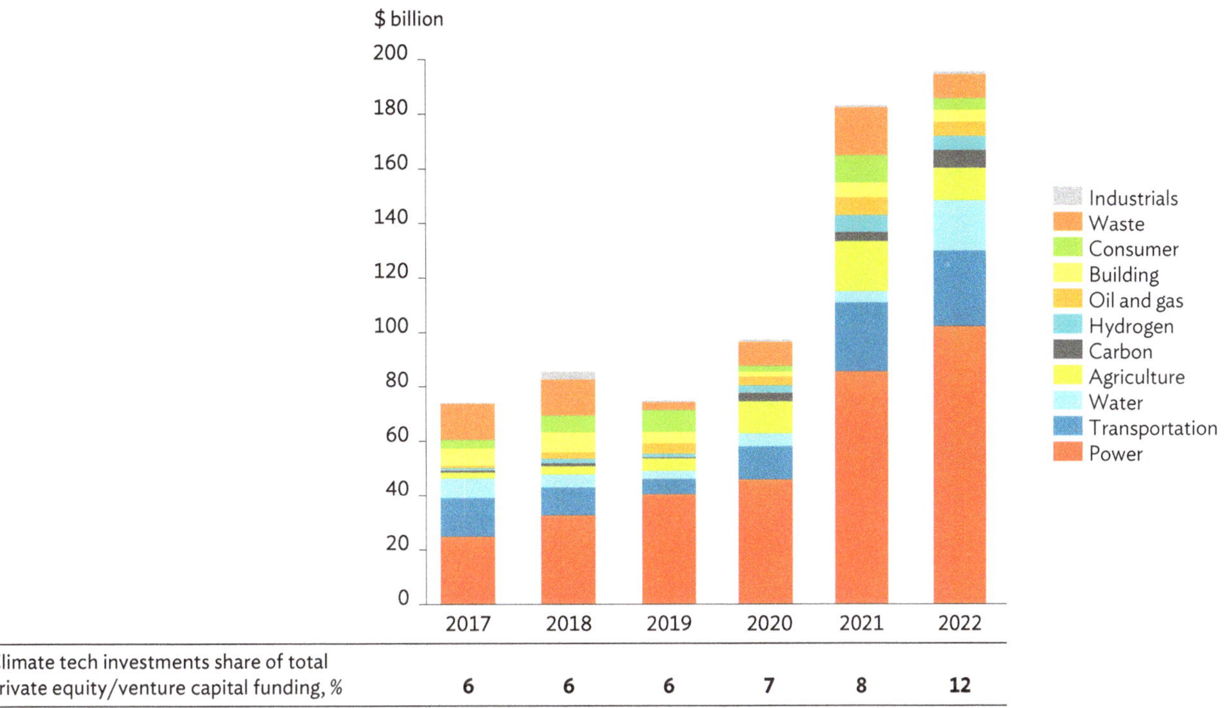

Source: Dahlqvist, F., S. Kane, L. Leinert, M. Moosburger, and A. Rasmussen. 2023. Climate Investing: Continuing Breakout Growth Through Uncertain Times. McKinsey & Company. 13 March.

The sustainable debt market drives opportunity in green and transition-related investments. Green, social, and sustainability (GSS) bonds as well as emerging bond labels such as sustainability-linked, blue, and transition bonds (known collectively as GSS+ bonds) saw notable growth from 2010 to 2022 (Figure 4.6). These products demonstrate strong potential to unlock private capital for climate action (J. P. Morgan 2023; Ehlers et al. 2022) and aim to deliver climate, socioeconomic, development, and nature co-benefits (Box 4.2). Figure 4.7 shows that Asia accounted for most sustainable debt issuances in 2023, totaling $140 billion. While the People's Republic of China (PRC) was responsible for about 85% of the 2023 total, the region's emerging-market and developing economies constituted less than 1% of Asia's sustainable debt issuance, mainly due to their limited access to international capital markets and limited domestic capital market development (Lim et al. 2024). Sovereign green bond issuances were an important catalytic factor in the development of private sustainable bond markets (Cheng, Ioannou, and Serafeim 2014).

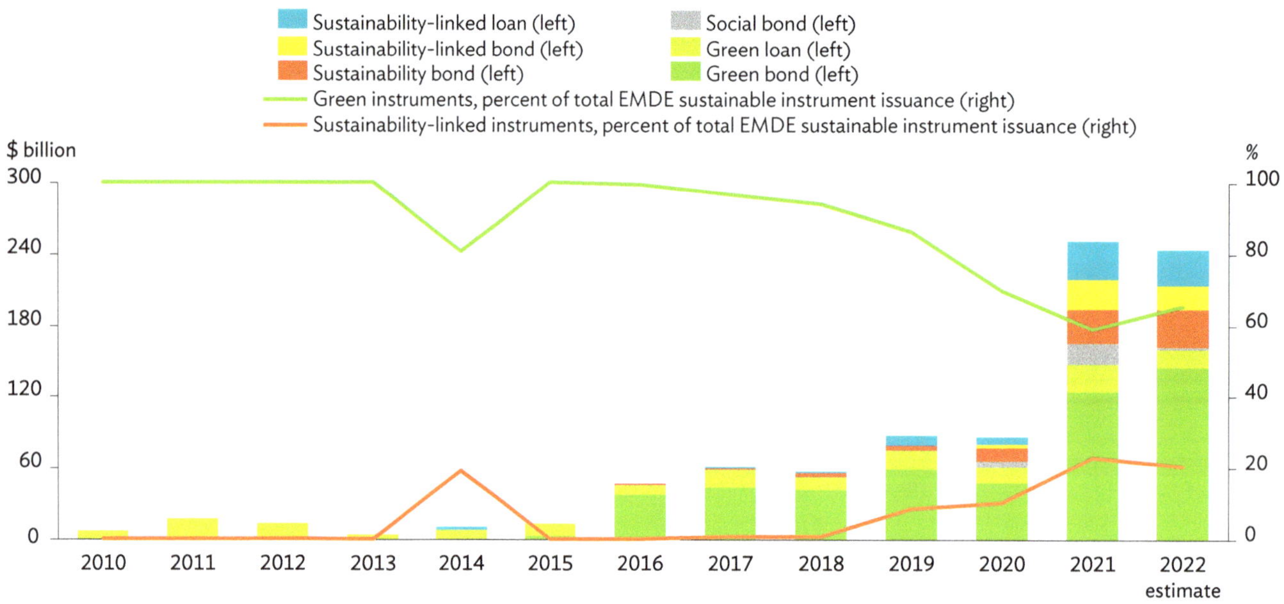

Figure 4.6: Sustainable Instrument Issuance in Emerging-Market and Developing Economies, by Instrument Type

EMDE = emerging-market and developing economy.
Source: Adapted from Figure 2.1 of Ehlers, T., C. Gardes-Landolfini, F. Natalucci, and P. Ananthakrishnan. 2022. How to Scale Up Private Climate Finance in Emerging Economies. *Global Financial Stability Report*. International Monetary Fund.

Figure 4.7: Sustainable Debt Issuance in Emerging-Market and Developing Economies, by Region and Economy Type

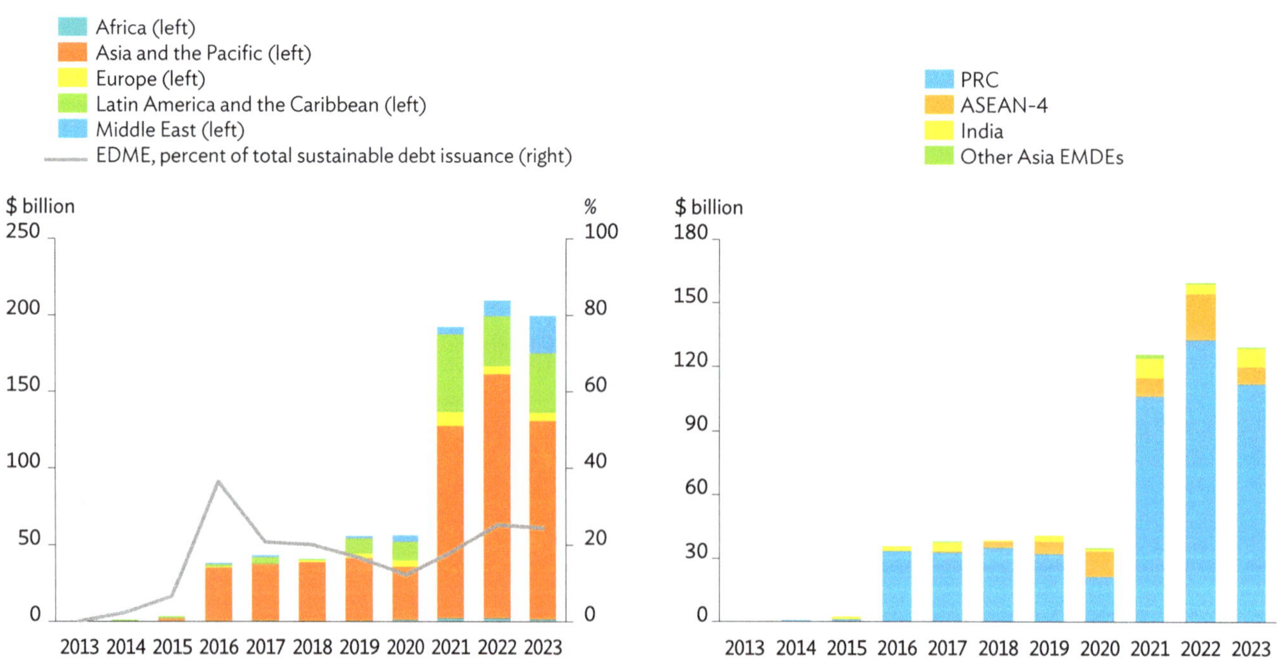

ASEAN-4 = Association of Southeast Asian Nations (Indonesia, Malaysia, Philippines, and Thailand); PRC = People's Republic of China; EMDE = emerging-market and developing economy.
Source: Asian Development Bank calculations based on International Institute of Finance data.

Box 4.2: Fiji's First Sovereign Blue Bond

In November 2023, with support from the United Nations Development Programme and the United Kingdom, Fiji announced the issuance of its first blue bond. This issuance seeks to raise private capital to unlock the potential of Fiji's blue economy through the sustainable use of its vast ocean resources. The bond will finance four priority areas with significant co-benefits:

(i)   using nature-based solutions for coastal protection of low-lying, vulnerable coastal communities;
(ii) scaling up Fiji's aquaculture sector by focusing on advanced hatchery programs for high-value species—such as sea cucumber, seaweed, grouper, and shrimp—to not only supplement imports but also support livelihood development in rural communities;
(iii) developing sustainable towns and cities using integrated planning solutions; and
(iv) enhancing solid waste management by building the capacity of Fiji's only sanitary landfill and boosting the capabilities of wastewater treatment.

Source: United Nations Development Programme. 2023. Launch of Fiji's First-Ever Sovereign Blue Bond. Press Release. 2 November.

Green and sustainable finance taxonomies help align investments with climate goals. A number of green or sustainable taxonomies have emerged in recent times. This has made information on what constitutes green and sustainable economic activities and investments increasingly available in the public domain (Figure 4.8). Such taxonomies function as comprehensive classification systems, defining eligibility for financing or investment based on the taxonomy's sustainability objectives (Wong 2024). In addition to providing frameworks for defining sustainable investments and activities, some taxonomies have made special provisions for transition finance, i.e., financing the decarbonization of high-emitting industries. Development in green and sustainable taxonomies is expected to provide clearer benchmarks to investors, thus helping channel financing toward eligible projects (Heinkel, Kraus, and Zechner 2001; Sharfman and Fernando 2008). While this clarity is likely to function as an incentive for increased financing of sustainable activities, associated disclosure requirements also act as regulatory push factor.

Figure 4.8: An Overview of Green and Sustainable Taxonomies, 2023

Existing	Under Development
• CBI Green Taxonomy • EU Taxonomy of Sustainable Activities • Association of Southeast Asian Nations • Thailand (Green and Transition) • Bangladesh Sustainable Finance Policy • PRC Green Bond Endorsed Projects Catalogue • Colombia Green Taxonomy • Georgia Sustainable Finance Taxonomy • Indonesia Green Taxonomy 1.0 • Malaysia Climate Change and Principle-Based Taxonomy • Mexico Sustainable Taxonomy • Mongolian Green Taxonomy • Russian Green Taxonomy • South Africa Green Finance Taxonomy • Korean Green Taxonomy (K-Taxonomy) • Sri Lanka Green Finance Taxonomy • Viet Nam Green Taxonomy • Kyrgyz Green Taxonomy on Sustainable Development	• Australia (Sustainable) • Canada (Green and Transition) • Chile (Sustainable or Green) • India (Green) • Japan (Green and Transition) • Kazakhstan (Green) • New Zealand (Climate) • Singapore (Green and Transition) • United Kingdom (Green) • Peru (Green) • Hong Kong, China (Green) • Dominican Republic (Green) • Panama (Green) • Costa Rica (Green) • Senegal (Green) • Rwanda (Green)

PRC = People's Republic of China, CBI = Climate Bonds Initiative, EU = European Union.
Source: Adapted from Azoulay, O., C. Merle, and L. de Souza. 2023. The New Geography of Taxonomies. Natixis.

Returns on green assets often outperform those on brown assets as climate concerns increase.
Such evidence has been documented in Pástor, Stambaugh, and Taylor (2021); Choi, Gao, and Jiang (2020),
Engle et al. (2020); and Ardia et al. (2021), among others. For instance, Ando et al. (2023) price the premium
for green sovereign bonds, the so-called "greenium", at about 11 basis points on average for emerging markets,
with the premium growing after issuance. However, the climate risk premium may not be accurately priced by the
market, possibly threatening financial stability (IMF 2020; Ehlers, Packer, and de Greiff 2021; Atz et al. 2022).
The profitability of climate-related investments can be improved when climate risks are better priced in the
financial system. Another inefficiency relates to climate risks shifting from more-transparent to less-transparent
market segments, slowing down the transition to net zero. For instance, Beyene et al. (2021) find that bond
markets price the risk of assets held by fossil fuel firms becoming stranded, while banks in the syndicated
loan market seemingly do not price this risk much. As a result, to meet their financing needs, fossil fuel firms
increasingly rely less on bonds and more on loans from banks.

A mix of climate policy and financial regulation helps mobilize private climate capital. Economy-level
climate commitments, such as NDCs and related laws, set climate goals and strategies. They guide expectations
for private investments and help reduce policy uncertainty. Some climate policies are in the form of command
and control (e.g., sector regulations, climate-aligned subsidies, and tax treatments), market mechanisms
(e.g., carbon taxes, emissions trading systems), and green nudging schemes (He 2023). These climate policies
create incentives for climate investments by improving their risk–return profile and shaping behavioral changes
in the marketplace (Table 4.1).

Table 4.1: Climate Policies and Private Climate Capital

Climate Policy	Correct Prices and Foster Market Responses
Climate Commitments	A. Reducing climate investment uncertainty • Comprehensive climate strategies, commitments, and transition plans • Climate laws
Standards, Regulations, and Subsidies	B. Supporting profitability and demand for climate solutions • Sector regulations and standards • Subsidies, green off-take, tax treatment, purchasing agreement, among others
Carbon Pricing	C. Pricing the externalities of carbon emissions • Carbon taxes • Comprehensive cap and trade (emissions trading scheme) • Removal of fossil fuel subsidies • Carbon Border Adjustment Mechanism
Green Nudges	D. Encouraging climate action with information • Green default settings • Eco and efficiency labeling highlighting costs and benefits to conservation to consumers
Financial Regulation	**Develop Climate-Oriented Financial Markets**
Monetary and Banking Regulations	E. Mainstreaming integration of climate risks • Requiring climate integration in lending and investment • Climate stress tests • Climate considerations for collateral
Sustainable Financial Market Development	F. Building a climate-oriented market ecosystem • Green and transition taxonomies and roadmaps • Sovereign issuance of green and sustainability bonds • Green and sustainable bond standards and principles • Capacity building for external green bond verification, certification, and rating
Climate Standards and Disclosure	G. Enabling transparency and integrity • Reporting standards (TCFD, ISSB) • Measures to support market integrity for climate investments

ISSB = International Sustainability Standards Board, TCFD = Task Force on Climate-Related Financial Disclosures.
Source: Mortimer, J. and S. Tian. 2024. Mobilizing Private Climate Finance – The Role of Climate Policies. Background paper for the *Asia–Pacific Climate Report 2024: Catalyzing Finance and Policy Solutions*. Asian Development Bank.

Effective policy can improve the profitability and commercial viability of climate-oriented projects.
Climate policies can directly boost the commercial viability of climate-related projects by creating market
incentives and stimulating market demand. Typical policy tools include imposing standards and regulations,
applying subsidies and tax incentives, and establishing binding carbon prices. Carbon prices can increase
economic returns in low-carbon investments by internalizing the price of emissions.[6] For instance, Hengge,
Panizza, and Varghese (2023) find that policies increasing carbon prices are effective in raising the cost of
capital for emissions-intensive firms. Command-and-control type standards and regulations prioritize low
carbon technologies, also phasing out emitting activities by reducing their scale and demand. Subsidies and
tax treatments (e.g., electric vehicle subsidies) promote low-carbon technologies to make those innovations
competitive with polluting activities. They also mitigate risks associated with low-carbon technologies
by facilitating long-term contracts and guaranteed prices such as feed-in tariffs (Koerner et al. 2022;
Roslan et al. 2022; Setiawan et al. 2022).

Public sector support is critical to help de-risk climate adaptation and mitigation projects. Public
sector funding of climate projects in developing economies has been deployed in the form of blended
finance. Instruments such as first-loss guarantees, subordinated loans, or hedging can offset investor fears
about unstable or uncertain project returns (Songwe, Stern, and Bhattacharya 2022; Ehlers et al. 2022).
Performance-based incentives for project developers can compensate for the marginal cost of using "greener"
technologies. This creates opportunities for private investors to limit risks while maximizing positive societal
impacts (WEF 2022; Dahlqvist et al. 2023). For example, the Green Climate Fund under the United Nations
Framework Convention on Climate Change, through taking a first-loss position in climate projects, mobilizes
four times the capital provided as public finance (WEF 2022).

**Economies with favorable investment conditions are more likely to attract foreign private capital for
climate action.** Besides stable macroeconomic conditions, international investors are attracted by strong
legal and regulatory frameworks for climate action. This includes adequate enforcement capacity, strong
institutions, effective and efficient public budget management systems (including for climate initiatives),
and clearly articulated climate investment priorities (Morita and Pak 2018). Emerging-market and developing
economies with ambitious NDCs that have priority sectors clearly defined; bankable, large-scale project
pipelines, notably in the clean energy sector; and efficient project implementation are big draws for external
financiers (Bowman 2022; IFC 2017; IMF 2023; IEA and IFC 2023).

Foreign investment into climate action responds both to domestic policies and those set abroad.
Around the world, industrial policies are deployed to accelerate the transition to net zero. For instance, the
Government of the United States' Inflation Reduction Act, 2022 is widely seen as the most ambitious climate
policy in the country's history. As outlined in Box 4.3, research highlights that raising foreign private climate
capital for Asian economies is subject to spillovers from the act. In addition, the Asia and Pacific region can
attract and retain foreign private climate capital through appropriate policies to reduce emissions, signaling
its strong commitment and capacity to tackle climate change. Moreover, policies to tilt the allocation of
conventional investment funds toward sustainability objectives can considerably increase the supply of
private climate capital.

---

[6]   Chapter 5 of this report provides a comprehensive review of the introduction of carbon pricing in the Asia and Pacific region and
discusses how carbon pricing creates incentives to mobilize climate finance.

Box 4.3: How the Inflation Reduction Act Increased Climate Finance in Asia and the Pacific

The United States' Inflation Reduction Act (IRA) aims to reduce the country's emissions by 40% (relative to 2005 levels) by 2030. A combination of tax credits, grants, and loans worth at least $370 billion promises to accelerate the transition to net zero by encouraging private sector investments in clean energy, including clean electricity generation and transmission, carbon capture and storage, and "green" hydrogen. Targeted subsidies stimulate demand for energy-efficient appliances, electric vehicles, rooftop solar panels, geothermal heating systems, and home batteries.

Research shows that the IRA has increased the supply of climate finance globally, including in Asia and the Pacific (te Kaat, Raabe, and Tian forthcoming). Investment funds are an important conduit for this spillover. Specifically, sustainable investment funds not domiciled in the United States received more inflows from investors after the announcement of the IRA on 27 September 2021. In turn, these sustainable funds increased their cross-border portfolio investments in recipient economies in Asia and the Pacific. Economies better prepared to address climate change benefited most from this windfall of additional investment inflows.

The analysis highlights the role of investment funds in facilitating an international spillover of green industrial policies. These fund flows were particularly prominent in Asia, cumulatively reaching about 30% of gross domestic product (GDP) on average from 2015 to 2022, with average quarterly growth rates of 16.5%. Over the same period, the People's Republic of China; Japan; Taipei,China; Australia; and the Republic of Korea together accounted for 90% of sustainable fund flows into Asia, as shown in the figure below.

Investment Fund Inflows for Selected Asian Economies as a Share of Total Regional Flows, 2015–2022
(average, % share)

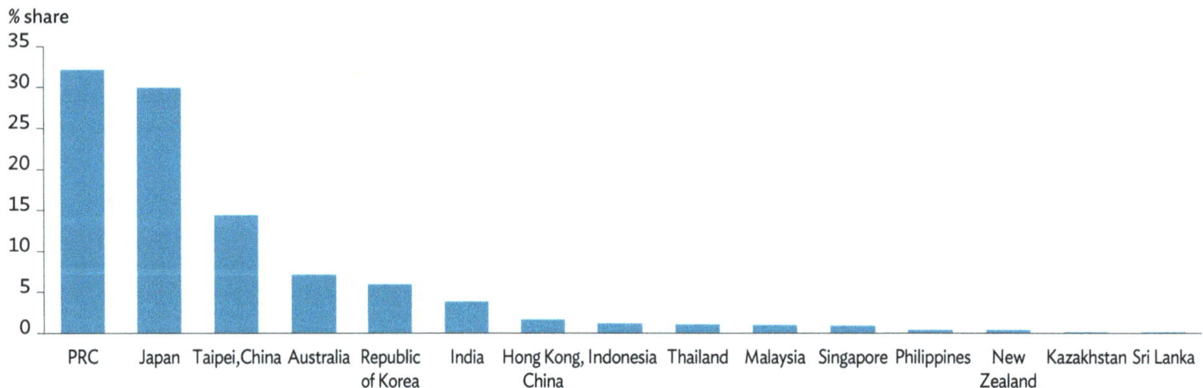

PRC = People's Republic of China.
Source: te Kaat, D. M., A. Raabe, and Y. Tian. Forthcoming. Greening Thy Neighbor: How the US Inflation Reduction Act Drives Climate Finance Globally. Asian Development Bank Working Paper.

The research suggests an increase in investments into sustainable funds relative to conventional funds rising by between 0.033 and 0.045 percentage points after the IRA announcement. This is equivalent to a 100% increase relative to the mean of sustainable fund inflows, or an additional $1.4 billion 3 months post IRA announcement. Given low and time-invariant cash shares in fund portfolios, additional inflows into funds translate directly into higher cross-border fund investments in recipient economies. For the median Asian economy, this implies additional sustainable investments cumulatively for 3 months after the IRA announcement equal to 3% of GDP on average as seen in the following Figure a.

continued on next page

Box 4.3 *continued*

Impact of Inflation Reduction Act and Recipient Economy Climate Policies

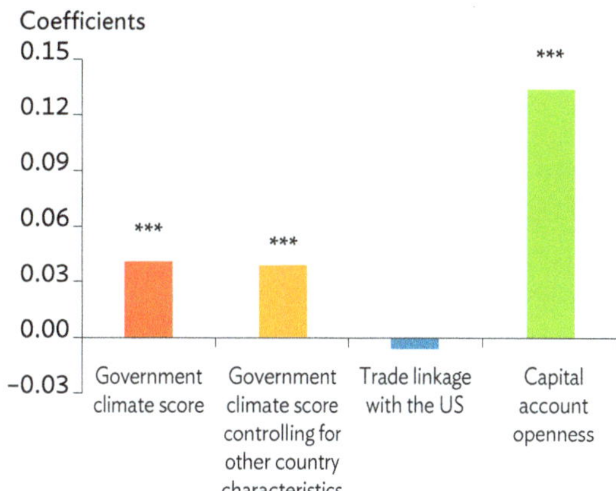

m = month, US = United States.
Notes: Figure a plots coefficients from a regression of flows into investment funds on the interaction between time dummies and a fund-level sustainability indicator, also applying appropriate controls. Figure b shows the coefficients from a regression of investment fund flows on a triple interaction comprising a post-Inflation Reduction Act dummy, a fund-level sustainability indicator as well as the Bloomberg Government Climate Score, or an economy's trade linkage with the United States, or an economy's financial openness. Stars denote statistical significance levels at 1%. The red vertical line denotes the IRA announcement, and blue vertical lines the confidence interval of coefficient estimates.
Source: te Kaat, D. M., A. Raabe, and Y. Tian. Forthcoming. Greening Thy Neighbor: How the US Inflation Reduction Act Drives Climate Finance Globally. Asian Development Bank Working Paper.

The research further investigates how authorities can harness these spillovers to attract more climate finance from abroad. Given large climate financing needs, it is important to allocate additional sustainable investments into those recipient economies where funds will be most productively channeled into climate action. An economy's capacity to productively use inflows for investments related to climate change can be proxied by Bloomberg Government Climate Scores. These scores measure an economy's progress in emissions reduction, power sector decarbonization, and policy commitments to address climate change. The analysis reveals that higher-scoring economies tended to attract more inflows from sustainable investment funds after announcement of the IRA, cumulatively worth an additional 1% of GDP for the average global economy and 3% on average for economies in Asia and the Pacific, as set out in Figure b above.

Additional investments were not only driven by higher inflows into sustainable funds investing in higher-scoring economies but also by an active portfolio shift of conventional funds in favor of economies with better climate policies. This suggests that conventional investment funds are an important source of climate finance, as they represent 97% of funds by assets under management.

Reference: te Kaat, D. M., A. Raabe, and Y. Tian. Forthcoming. Greening Thy Neighbor: How the US Inflation Reduction Act Drives Climate Finance Globally. Asian Development Bank Working Paper.

Source: Authors.

4.3.1.2 Push Factors

Managing disruptions caused by extreme weather events has motivated firms to invest in mitigation and adaptation. While this scenario is widely documented in relevant literature (e.g., Cox et al. 2022; Randall, Sedemund, and Bartz-Zuccada 2023; Flynn et al. 2023), such business strategies may not always be labeled as climate investments. For example, a firm's general expenditure on research and development may reflect higher vulnerability to climate change (Alam et al. 2023). Similarly, real estate investment trusts have been investing in energy efficiency, GHG reduction, and water resource management strategies (Majersik 2019).

Policy and regulatory initiatives are a key push factor for private climate investments. Akomea-Frimpong et al. (2022) identify global action on climate change as the top determinant for green finance by banks. Regulatory frameworks mandating emissions reductions, climate risk assessments, and climate-related financial risk disclosure have been introduced in ever more jurisdictions and play an increasingly significant role in driving private sector investments in climate action (EU 2022; BEIS 2022; ERB 2022; HKEX 2023; ISSB 2023; Arca and SGX 2024). Box 4.4 provides details on a 2023 financial disclosure regulation by the European Union (EU). The introduction of the EU's carbon border adjustment mechanism may induce Asian exporters' compliance with EU emission standards in an effort to avoid export declines (ADB 2024).

Box 4.4: The European Union Sustainable Finance Disclosure Regulation

The European Union's Sustainable Finance Disclosure Regulation (SFDR), enacted in February 2023, requires asset managers and other financial market participants to disclose the extent to which investments are aligned with environmental, social, and governance criteria. The regulation applies to all funds operating in Europe. Under the SFDR classification system, investment funds fall into one of three categories: Article 6 (no sustainability focus), Article 8 ("light green", promoting environmental characteristics), or Article 9 ("dark green", a clear objective of sustainable investment). Evidence suggests that the introduction of the SFDR has enhanced transparency, with more capital being channeled toward verified sustainable investments.

Source: International Monetary Fund. 2023. Global Financial Stability Report: Financial and Climate Policies for a High-Interest-Rate Era.

Demands of key stakeholders constitute another important push factor. Such evidence has been documented by Kawabata (2019) and Diaz-Rainey et al. (2023), among others. For example, large asset owners, including Japan's Government Pension Investment Fund (the world's biggest pension fund) are mandating that their fund managers delve into the environmental, social, and governance (ESG) credentials of the companies in which they invest (The Economist 2020). Pressure has also grown for private equity firms to integrate climate-related risks into their investment processes, guiding these firms toward alignment with broader climate goals and initiatives (ERM and Ceres 2021).

4.3.2 Indirect Financial Drivers

Some indirect financial factors have been influential in driving private climate investment. Such factors motivate investment in climate action without the expectation of immediate or direct financial return. They can, however, provide indirect financial gain in the longer term. For example, by investing in coastal protection programs, a multinational tourism resort may gain the support of local communities, which may strengthen the firm's brand and result in subsequent financial advantage.

4.3.2.1 Pull Factors

Ethical and sustainable investments, particularly ESG investing, have significant pull for investors.
Many institutional investors, as well as individuals with high net worth, are pivoting toward investments in
line with ethical values. Institutional investors, such as pension funds, own a large proportion of global assets
and are therefore highly exposed to climate risks (J.P. Morgan 2023). ESG investing serves as a framework
to incorporate sustainability values, resulting in higher long-term financial value (Eccles 2021). Similarly,
investments by high net-worth individuals are increasingly guided by environmental and social benefits,
beyond financial returns, and this is especially so in Asia. For instance, on average in 2024, 72% of young
high net-worth individuals across Asia factored sustainability concerns into investment decisions, compared
to 43% in Canada, the United Kingdom, and the United States (Deloitte 2024). Despite a temporary dip
in 2022–2023, global ESG investment remained strong in early 2024 (Temple-West and Masters 2024).
With the lion's share of ESG flows still going to funds domiciled in Europe, ESG flows and their market share in
Asia and the Pacific saw steady growth from 2016 to 2022, as shown in Figure 4.9 (J.P. Morgan 2023).

Figure 4.9: Market Share of Environmental, Social, and Governance Funds, by Region

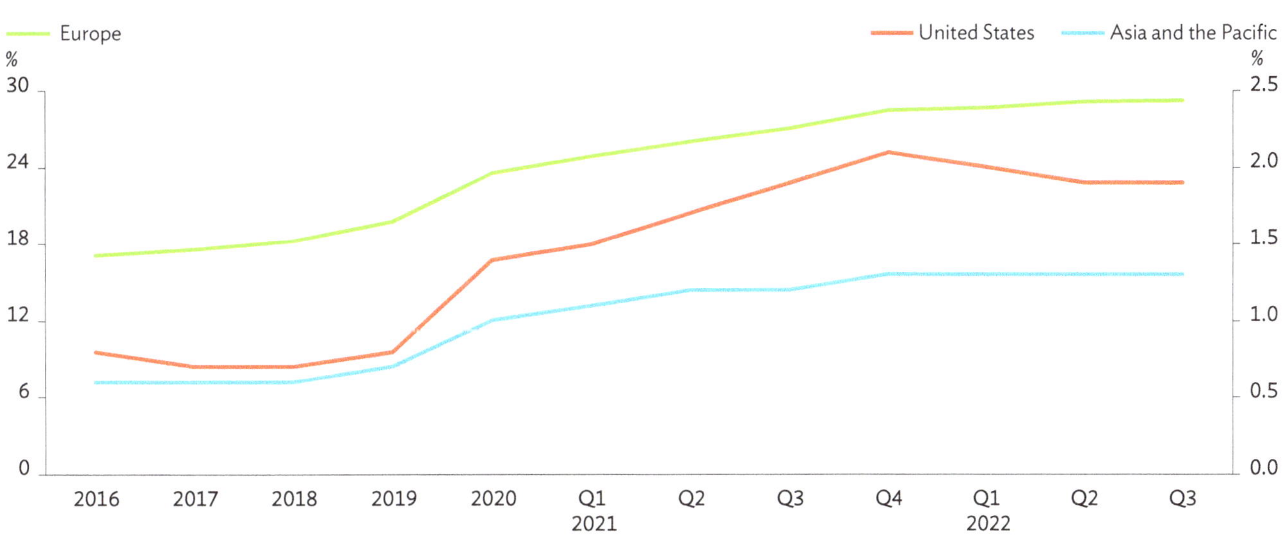

Q = quarter.
Note: Percentages are based on assets under management.
Source: Adapted from Figure 35 of J. P. Morgan. 2023. Asia Pacific Equity Research. 23 January.

Another pull factor is the potential for brand and corporate image enhancement. Environmental
responsibility by commercial entities is increasingly shaping consumer preferences and investor perceptions
(Henderson, Ghosh, and Zhan, 2018; Du, Wenxiu, and Zhang 2020). Companies that demonstrate a
commitment to sustainability therefore gain a competitive advantage, including increased market share,
brand loyalty, and access to capital. An ADB survey conducted in 2022 confirms that investment strategies
integrating Sustainable Development Goals contribute to an organization's "greener" image (ADB 2022).

4.3.2.2 Push Factors

Customer preferences and/or general public opinion can accelerate investment decisions. The general public, particularly younger generations, tend to be loyal to corporates acting on climate change (Robins, Brunsting, and Wood 2018; SustainAbility Institute 2021). This is particularly important when the labor market is tight and competition for young talent is strong (e.g., Weinert 2023). The Deloitte (2024) survey confirms that environmental sustainability remained the top priority of young people and that it consistently weighs heavily on them, with roughly 6 in 10 respondents saying they had felt worried or anxious about climate change. Young people therefore tend to make purchasing decisions based on environmental considerations (Figure 4.10a). Moreover, more than half of all respondents in the Deloitte survey reported that they have attempted to influence their employers to take action on climate change (Figure 4.10b).

Figure 4.10: Consumption and Career Choices Made by Younger Generations, 2024

(a) Consumption Patterns

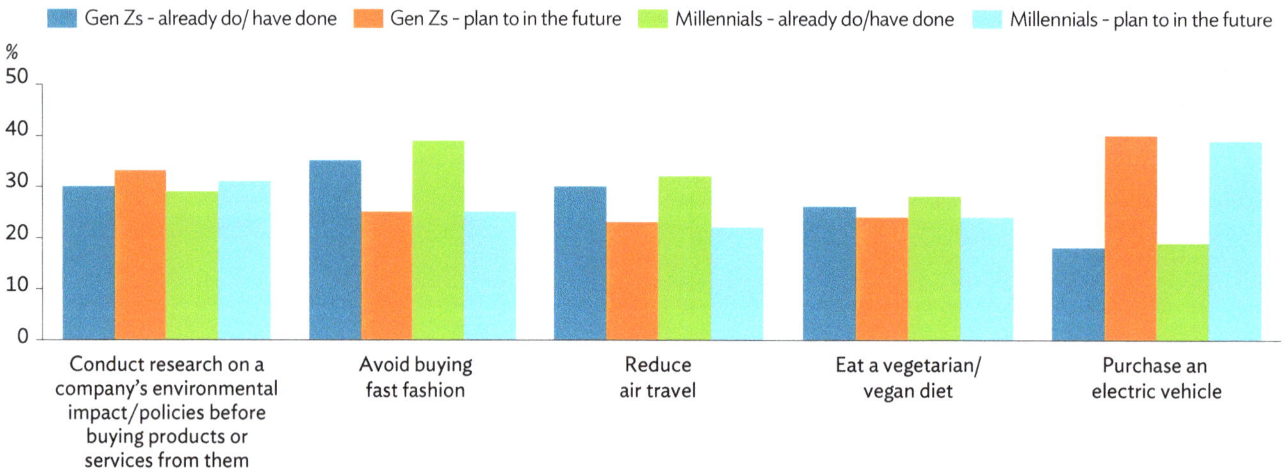

(b) Career Decisions

	Having already changed or planning to change job or industry due to climate concerns	Saying they and their colleagues are putting pressure on their employers to take action on climate change
Gen Zs	46%	54%
Millennials	42%	48%

Note: "Gen Z" is defined as those born after 1997, while "Millennials" are born between 1981 and 1996.
Source: Deloitte. 2024. 2024 Gen Z and Millennial Survey: Living and Working with Purpose in a Transforming World.

Climate finance initiatives can help commit corporates to climate-related policies and investments. Employee awareness on climate change and management-level engagement in climate-related programs are associated with a higher level of climate finance engagement by financial institutions (Kawabata 2019). When a firm commits to membership of an international climate finance initiative, it is expected to abide by the initiatives' investment parameters and targets, ultimately establishing a degree of "climate credibility" in the public eye.[7]

Climate-related litigation has triggered commercial entities to act on climate change. Climate-related litigation has grown rapidly (NGFS 2023). Companies in energy generation, manufacturing, transport, and agriculture have been targeted for a range of claims. These claims include damages for past emissions, violation of corporate due diligence laws, and allegations of "greenwashing" (i.e., investments overstating climate benefits). Financial institutions have also been accused of greenwashing and violation of due diligence laws (NGFS 2023).[8] Litigation concerning investment decisions has been increasing and can help clarify the parameters for such decisions in the context of climate change (Setzer and Higham 2023; ADB 2020).

4.4　Addressing Impediments to Mobilizing Private Climate Capital

Recent developments in climate policies, financial markets, and marketplace responses have mobilized private capital toward climate-oriented investments. Nevertheless, impediments remain, including fragmented climate policy implementation, underdeveloped climate-aligned financial markets, and marketplace frictions.

4.4.1　Consistent and Aligned Climate Policies

4.4.1.1　Challenges to Climate Policy Implementation

Uncertainty in climate policy hinders private investment by raising risk perception. Political considerations are known to cast uncertainty on climate policies. Some governments choose to temporarily halt or soften implementation of climate policies due to other commitments or political priorities. In the United States, the index of climate policy uncertainty increased from less than 10 in March 2000 to more than 250 in December 2021 (Figure 4.11). Europe has also witnessed opposition to green policies that constrain emissions and phaseout fossil fuels. This is largely due to the high costs placed on local businesses, farmers, and households. Such public opposition has placed pressure on a few European governments—including those in Germany, Italy, the Netherlands, Poland, Spain, and the United Kingdom—to soften some climate policies by delaying implementation or carving out exemptions. Additionally, disruptions to energy supplies due to geopolitical tensions, such as the Russian invasion of Ukraine and the closure of the Suez Canal, have redoubled government attention on energy security. As an example, Norway approved 19 new oil and gas projects in June 2023 to reduce its dependence on imported oil (Treloar 2023). This uncertainty around climate policies could lead firms to delay climate-related investments and reduce financing of new equipment and research and development (Bloom 2009). Therefore, greater certainty in climate policies and commitments can foster increased private climate investment via consistent expectations of future cash flows and risks.

---

[7]　Examples of such initiatives are the United Nations Principles for Responsible Investment, the Institutional Investors Group on Climate Change, the Investors Group on Climate Change, the Asia Investors Group on Climate Change, the Ceres Investor Network on Climate Risk, the Investment Leaders Group, the RE 100 Campaign, the We Mean Business Coalition, and Climate Action 100+.

[8]　The Sabin Centre for Climate Change Law maintains a Climate Change Litigation Databases of worldwide climate-related litigation cases and can provide further details on each case.

Figure 4.11: Climate Policy Uncertainty Index in the United States

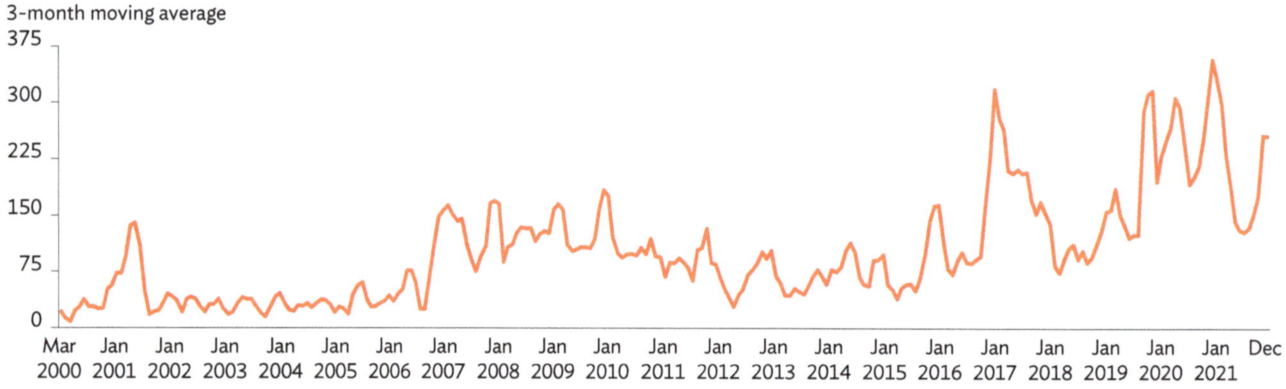

3-month moving average

Sources: Economic Policy Uncertainty (accessed 17 June 2024); and Gavriilidis, K. 2021. Measuring Climate Policy Uncertainty. 16 May.

Lack of policy alignment weakens incentives for private climate investors. In assessing competing budget priorities, governments must balance the need for economic growth, social inclusion, and environmental protection. When treated in isolation, policies to achieve these goals can become fragmented and sometimes contradict each other, thereby reducing incentives designed to stimulate private climate investment. For example, many governments in developing Asia are subsidizing renewable energy investment and have introduced carbon pricing, but they continue to spend significant amounts on fossil fuel subsidies (ADB 2023d). Fossil fuel subsidies weaken the effect of carbon pricing and renewable energy subsidies, both through the reduced competitiveness of low-carbon projects and the mixed signals sent to private investors. Specific climate policies, especially when they come from different authorities, may also not work with each other in a consistent or complementary manner (He 2023). For instance, the effectiveness of carbon pricing can be reduced when regulations or standards (e.g., energy efficiency standards, renewable portfolio standards) are imposed on certain firms or industries. This is because firms will prioritize meeting standards or regulations before trading carbon allowances, in turn lowering demand for carbon allowances. This reduced demand for carbon credits then translates into lower carbon prices and reduced incentives.

4.4.1.2 Pursuing Aligned Climate Policies to Mobilize Private Climate Capital

Policymakers can incentivize private climate investments by reducing policy uncertainty. Investors consistently state that a stable and predictable policy environment is one of the most important factors in long-term investment attractiveness (BCG 2023; Olasehinde-Williams, Özkan, and Akadiri 2023; ORF 2023). Ciminelli et al. (forthcoming) demonstrate that greater climate policy certainty, achieved through the legislation of climate commitments as law, helps boost sustainable investment (Box 4.5). The public sector can also reduce investor perception of climate policy risk by announcing climate commitments backed with practical measures (such as sectoral and interim targets) and financing strategies. For example, when economies set out a net zero pledge, national green development policy, or strategic green growth plan, corresponding green industrial policies or climate investment programs will signal greater certainty about the ability to achieve climate goals. Since green industrial policies or climate investment programs do not directly draw on public finances, they can cost-effectively stimulate private climate capital by lowering investment risk perceptions. Creating an enabling environment with legislation that addresses investor concerns would boost investor confidence overall. This requires political commitment over the medium to long term (Mik 2024).

Governments can adopt a comprehensive set of policy instruments to incentivize private climate capital. Nascimento and Höhne (2023) show that expanding climate policy adoption across sectors, with complementary price and nonprice measures, is effective at achieving emissions reductions. A comprehensive policy package is not just having more policy tools, but also ensuring a mixture of different types of policies, such as price and nonprice enabling measures. A comprehensive policy package will strengthen climate policy certainty and thus enhance the profitability and incentives of climate-related investments, encouraging more private capital. The energy sector serves as an example. Besides subsidies and feed-in tariffs, auctions (as a market reform) can further efficiently incentivize investment projects in the renewable energy sector through green offtake mechanisms. Green offtake mechanisms provide investors with clarity on future demand, thereby improving commercial viability. Other examples include on-grid prioritization for renewable energy generation, mandates, and standards. Figure 4.12 indicates that climate policy adoption by economies in developing Asia is less comprehensive than in advanced economies.

Box 4.5: Climate Law and Sustainable Fund Holdings

Since the late 2000s, over 30 economies around the world have enshrined into law ambitious, economy-wide emissions reduction targets. Adopting a climate law is distinct from implementing a climate change mitigation policy because, while both aim to reduce greenhouse gas emissions, a climate law serves to bind all current and future government policies.

Adopting a climate law increases government accountability, offering legal recourse to firms and citizens in case of inaction. For instance, the Government of the United Kingdom's plans for hitting climate goals set in the government's Climate Change Act have twice been ruled unlawful by the United Kingdom's High Court. In this sense, climate laws establish a proper legal framework for economies to tackle climate change.

To the extent that investors reward legal and policy certainty, climate laws can play an important role in generating green finance. Climate laws provide greater certainty surrounding both an economy's commitment to the low-carbon transition and the stringency of future climate change mitigation policies. This creates a predictable environment for investors, encouraging them to invest in green technologies.

Empirical evidence supports the catalytic role of climate laws in eliciting green finance. Using a difference-in-differences regression analysis, economies that adopt climate laws see steady increases in the share of their assets held by environmental, social, and governance (ESG) funds over total assets held by mutual funds after approval, relative to economies without climate laws. This increase is statistically and economically significant. Specifically, the share of assets held by ESG funds increases by an average of about 2 percentage points (almost half of the sample average) 12 quarters after the approval of a climate law, and continues growing over time, as shown on the left side of the figure below. This dynamic is particularly pronounced among the subset of ESG actively managed funds and bond funds, as measured on the right side of the figure.

continued on next page

Box 4.5 *continued*

Climate Laws and Environmental, Social, and Governance Funds Asset Holdings

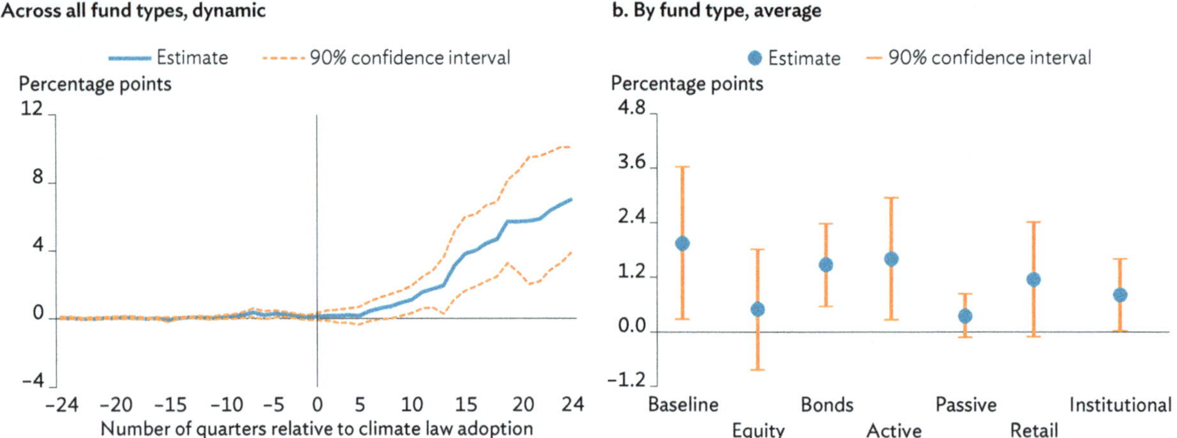

Notes: The two graphs show the differences between economies that adopt a climate law and those that do not, in terms of the share of their bond and equity assets held by environmental, social, and governance (ESG) funds over total assets held by mutual funds. Panel "a" depicts this difference (y-axis, percentage points) over time (x-axis, in quarters relative to adoption). Panel "b" reports the average difference after climate law adoption relative to the average difference before adoption by fund type (x-axis).
Source: Asian Development Bank calculations based on Ciminelli et al. Forthcoming. Climate Law and Sustainable Fund Holdings.

Source: Ciminelli et al. Forthcoming. Climate Law and Sustainable Fund Holdings.

Governments can strengthen incentives and enable private climate capital via an aligned and holistic policy package. Aligning policy steps, such as removing fuel subsidies can strengthen incentives without consuming extra fiscal resources, but this requires aligned planning and coordinated action across government departments. Governments can deliver a holistic policy package to strengthen incentives for private climate capital without extra spending. They can do so by pairing climate policies and regulations with enabling infrastructure and supporting systems, which contemporaneously matches supply and demand and enables private investment. Figure 4.13 shows examples of holistic policy packages that address system bottlenecks by pairing sector-specific policies with support for enabling infrastructure.

Figure 4.12: Regulation and Standards Versus Subsidies in "Green" Industries

Legend (Less ←—Implementation—→ More):

	Renewable Energy					Green Buildings		Clean Transportation	
	Auctions/ Tenders (Utility Scale)	Auctions/ Tenders (Prevalence)	Feed In Tariff or Premium	Grid Prioritization	Mandates	Building Energy Efficiency Codes	Incentives Available	Standards or Mandates (Clean Transport Target)	ICE Technology Phaseout (National, Light Duty Vehicles)
More	In Force	Auctions-Multiple Renewable Energy Consistently	Ended	In Force	In Force	Mandatory/All	Yes	Yes	Yes
		Auctions-Multiple Renewable Energy				Mandatory/Some			
	Suspended	Auctions-1–2 Renewable Energy	In Force	Partially In Force	Partially In Force	Voluntary/All		Somewhat	Somewhat
		Auctions-Announced				In Development			
Less	No Policy	No Auctions Announced/ No Data	No Policy	No Policy	No Policy	No Known Code	No	No	No

Advanced economies

- United Kingdom
- Germany
- United States
- Japan

Developing Asian economies

- Azerbaijan
- Bangladesh
- Cambodia
- PRC
- India
- Indonesia
- Kazakhstan
- Republic of Korea
- Lao PDR
- Malaysia
- Mongolia
- Pakistan
- Philippines
- Singapore
- Sri Lanka
- Taipei,China
- Thailand
- Uzbekistan
- Viet Nam

PRC = People's Republic of China, ICE = internal combustion engine, Lao PDR = Lao People's Democratic Republic.
Source: Mortimer, J. and S. Tian. 2024. Mobilizing Private Climate Finance – The Role of Climate Policies. Background paper for the *Asia–Pacific Climate Report 2024: Catalyzing Finance and Policy Solutions*. Asian Development Bank.

Figure 4.13: Examples of Holistic Policy Packages

Address system bottlenecks by pairing sector-specific policies with adequate support for enabling infrastructure	Enhance market efficiency by matching supply-side (issuer) sustainable finance standards with demand side-side (investor) expectations
• Renewable energy (RE) capacity installation and generation targets paired with electrical grid infrastructure investment and permitting reform • Personal transportation emissions efficiency standards or mandates paired with Electric Vehicle (EV) charging infrastructure support and incentives	• Corporate climate disclosure requirements paired with investor stewardship codes and fund climate reporting to build end-user demand for such data • Green and transition finance taxonomy implementation paired with national green bond and sustainable financial product standards to match high-quality supply with demand

Source: Mortimer, J. and S. Tian. 2024. Mobilizing Private Climate Finance – The Role of Climate Policies. Background paper for the *Asia–Pacific Climate Report 2024: Catalyzing Finance and Policy Solutions.* Asian Development Bank.

4.4.2 Mobilizing Private Climate Capital via Enabling Financial Systems

4.4.2.1 Challenges for Finance Markets

Lack of reliable climate-related information significantly raises transaction and financing costs.
Financial markets can sustainably mobilize and generate huge amounts of private investment by efficiently allocating savings to climate-aligned projects. However, weak climate information architecture increases the risks of greenwashing and reduces market transparency (IMF 2023). Lack of reliable climate-related information thus exposes investors to higher transaction costs and reputational risks. Therefore, financiers enjoy lower financing costs on climate-aligned investments when they can justify greater climate credibility and quality information. For example, green bonds carrying a widely accepted third party certification save 24–36 basis points on their bond yields, compared to green bonds without such certification (Hyun, Park, and Tian 2021). The lack of disclosure for climate-related risks is an important factor in explaining the relatively low level of investment in climate adaptation and resilience-building (Tall et al. 2021). In developing Asia, financial markets are still evolving to reduce deficiencies in climate-related information to encourage private climate capital. Generally, emerging-market and developing economies lack high-quality, reliable, and comparable climate-related data, making the assessment of risks and opportunities more complex for private investors (IMF 2023). The absence of quality information and mandatory disclosure requirements partly make investors such as banks and insurance companies overlook climate financing needs (IMF 2023).

Climate disclosure practice in developing Asia has made good progress, but it should be applied more broadly. In some economies of the Asia and Pacific region, publicly listed companies are required to comply with climate disclosure benchmarks set by their respective stock exchanges. This is the case in the PRC; Hong Kong, China; India; Indonesia; Japan; Malaysia; and Singapore (GRI 2022).[9] However, such disclosure is generally confined to larger companies. Around 80% of the top 50 publicly listed companies in 14 Asian markets reported their scope 1 and 2 GHG emissions in 2022 (PWC 2023). Climate disclosure becomes less common when a broader range of firms is considered, such as companies not publicly listed, especially small

---

9   More information on climate disclosure benchmarks can be found at Climate Governance Initiative (2023); GRI (2022); and Bloomberg (2024).

and medium-sized enterprises which lack capacity and resources to report emissions yet are crucial to supply chains. The adoption of International Sustainability Standards Board (ISSB) standards helps regional institutions build a global comparable baseline of sustainability information, so that investors and clients can make informed decisions. While the adoption of ISSB standards remains at an early stage worldwide, developing Asia is making good efforts on adoption. According to Laidlaw (2024), based on ISSB implementation progress outlined by IFRS (2024) and Bangladesh Bank (2023), Bangladesh was the first Asian economy to formally adopt the ISSB standards through guidelines for financial institutions in 2024. Meanwhile, several Asian economies, including the PRC; Hong Kong, China; India; Japan; the Republic of Korea; Malaysia; Pakistan; the Philippines; Singapore; Sri Lanka; and Taipei,China, are at various stages of planning and preparation for adoption and alignment with domestic standards (Laidlaw 2024; IFRS 2024; Bangladesh Bank 2023). Association of Southeast Asian Nations economies, including Indonesia, Thailand, and Viet Nam, are engaging with the ISSB to explore ways of incorporating the standard's global baseline in the region (Mortimer and Tian 2024). Developing Asia can benefit from accelerating the adoption of climate disclosure, not only to reduce compliance costs of companies but also to become more competitive when carbon pricing becomes mainstreamed, especially with the implementation of the European Union's Carbon Border Adjustment Mechanism.

Sustainable finance remains a small fraction of financial markets, and the banking sector can play a larger role. Sustainable finance, green finance, and climate finance can differ in activity coverage and investment objective. According to Berrou, Dessertine, and Migliorelli (2019) and the China Council for International Cooperation on Environment and Development (2022), sustainable finance supports sustainable economic activities including environmental, social, and other Sustainable Development Goals in the investment process (Figure 4.14). Green finance refers to investments aimed at climate and other environmental (e.g., nature and biodiversity) goals. Meanwhile, climate finance refers to financing that is dedicated to supporting climate change mitigation and adaptation actions. While some climate finance statistics cannot directly take into account the sustainable debt markets due to different activity coverage, green, sustainability, and sustainability-linked debt instruments dominate the global private sector sustainable debt outstanding, accounting for 95.5% ($4.1 trillion) in June 2024. Despite its rapid expansion, the sustainable debt market remains a small portion of the total private debt market. By June 2024, sustainable corporate bonds outstanding in developing Asia totaled $477.9 billion, accounting for only 4.2% of the general corporate bond markets in the region. The corresponding share in the world's largest regional sustainable bond market, EU20, was higher at 7.8% due to investor demand and regulatory support, among other factors. While regional central banks are starting to introduce environmental and social frameworks to guide bank lending, it is crucial for the banking sector to play a larger role in financing sustainable projects and label related green and sustainability activities in the region. In selected developing Asian markets, where loan data were available, labeled sustainable loans (including green, social, sustainability, and sustainability-linked loans) accounted for around 35.3% of total corporate sustainable debt financing (including both loans and bond issuances) as of March 2024. This share is significantly lower than bank credit's share of 77.6% in total corporate debt financing, where corporate bonds accounted for only 22.4% as of March 2024.[10]

---

[10]   The statistics include selected economies of developing Asia due to availability of bank loan data. Markets included are the People's Republic of China; Hong Kong, China; India; Indonesia; the Republic of Korea; Malaysia; Singapore; and Thailand. Statistics were sourced from AsianBondsOnline computations based on Bank for International Settlements and Bloomberg LP data.

Figure 4.14: Sustainable Finance Definitions

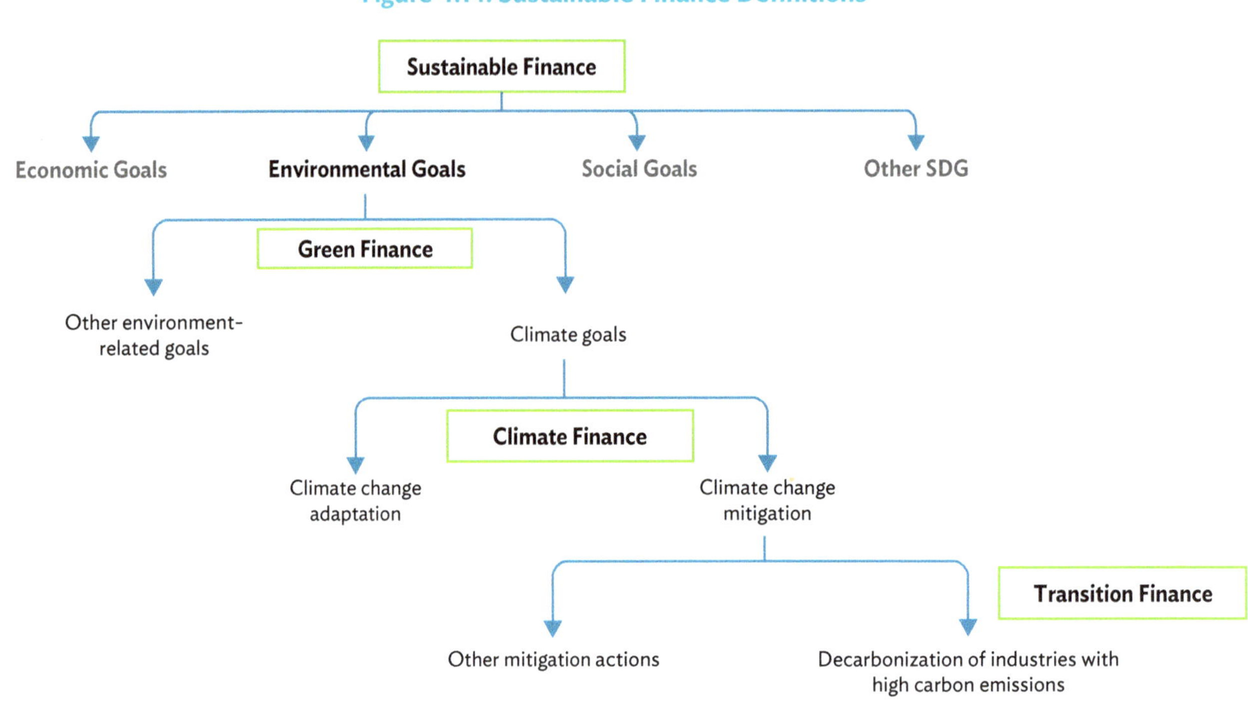

SDG = Sustainable Development Goal.
Sources: Asian Development Bank visualization based on Berrou, R., P. Dessertine, and M. Migliorelli. 2019. An Overview of Green Finance. In M. Migliorelli, and P. Dessertine, eds. *The Rise of Green Finance in Europe.* Palgrave Macmillan; and China Council for International Cooperation on Environment and Development. 2022. Green Consensus and High Quality Development.

More support is needed to expand the use of sustainable finance market in financing firms' decarbonization efforts. As shown in Figure 4.14, transition finance supports decarbonizing emissions-intensive industries and activities to meet net zero targets, making it an important aspect of climate finance.[11] Mobilizing credible transition finance helps address carbon lock-in and achieve national net zero targets. While supporting the productive transition of otherwise stranded assets, it thus contributes to financial stability (OECD 2023). Transition finance has emerged as a fast-growing area in sustainable bond markets. According to the International Capital Market Association (ICMA 2024), transition activities financed in the sustainable bond markets are captured by green, sustainability, and sustainability-linked bonds issued by the fossil fuel sector and hard-to-abate industries (such as steel, cement, shipping, and aviation), as well as sustainable bonds with a transition bond label. In developing Asia, the cumulative issuance of such labeled bonds reached $448.7 billion by June 2024, up from $35.2 billion in 2020, a 12.7-fold expansion that is largely in line with the 12.3-fold growth in the region's entire sustainable bond issuance over the same period. However, the share of transition finance instruments remains only a small fraction of sustainable bond markets, accounting for a mere 6.2% of the region's sustainable bonds issuance as of June 2024 (Figure 4.15). ICMA (2024) indicates that it is difficult for the fossil fuel sector and hard to abate industries to raise transition finance due to a lack of consensus on acceptable and credible technologies and trajectories. Furthermore, investors and issuers are often disincentivized from investing in brown-to-green projects by binary green finance regulations and the risk of accusations of greenwashing. EY Global Financial Services (2024) summarizes a few key impediments to transition finance, including a lack of credible and interoperable transition taxonomies to guide investments in decarbonization activities, a lack of sound transition planning to guide financial institutions' investments, a lack of economic viability in coal phaseout projects, and lack of demonstration of successful decarbonization projects.

---

[11] According to the International Capital Market Association (ICMA 2024), a broader definition of transition finance can support economy-wide transition.

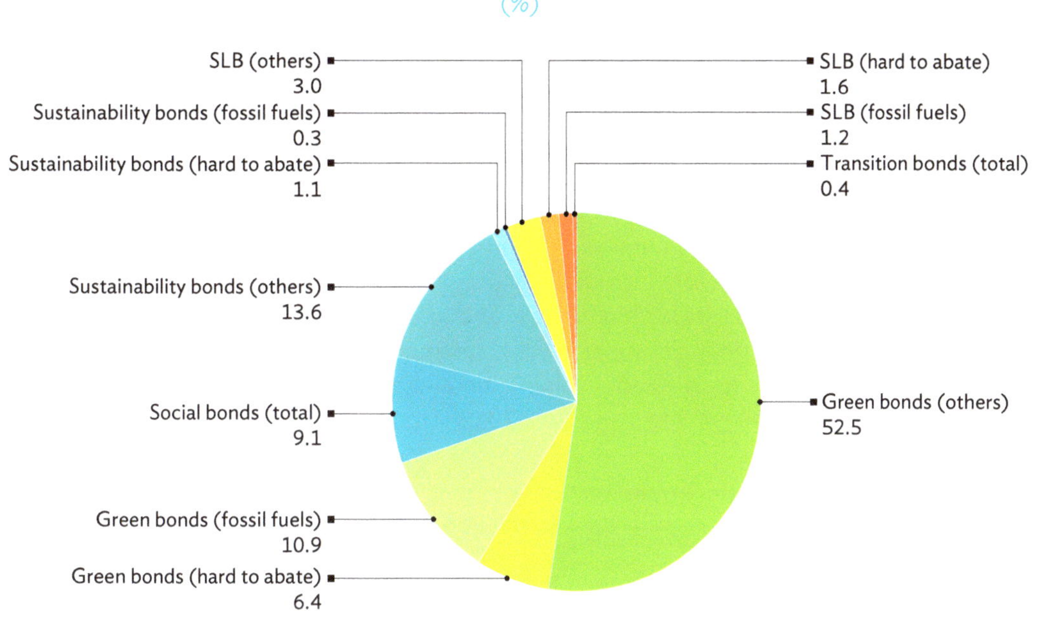

Figure 4.15: Sustainable Bonds to Support Transition Activities in Developing Asia
(%)

SLB = sustainability-linked bonds.
Notes: Data refer to cumulative issuance from 1 January 2019 to 30 June 2024. "Hard-to-abate" industries include airlines, chemicals, construction materials manufacturing, metals and mining, railroad, transportation and logistics, and utilities. "Fossil fuels" industries include coal operations, exploration and production, integrated oils, pipelines, and power generation.
Source: AsianBondsOnline calculations based on Bloomberg LP data.

Multiple sustainable finance taxonomies can be particularly challenging for cross-border flows of climate finance. Sustainable finance taxonomies are an important part of the financial market ecosystem as they provide standardized principles and classifications, including through the specification of assessment criteria for green and transition activities. As of March 2024, more than 50 official national and regional sustainable finance taxonomies had been developed or were in development (Wong 2024). These taxonomies may adopt different approaches and assessment criteria to account for country-specific economic and social circumstances and resources, under the principle of "common but differentiated responsibilities" in the Paris Agreement. Discrepancies across national and regional taxonomies may raise transaction costs in mobilizing international climate investments. For example, an activity structured to be eligible under one taxonomy may have to be reassessed, or may become ineligible, under a different taxonomy. Investments can therefore be disincentivized due to increased assessment and transaction costs and extended processing periods. Multiple taxonomies can also disrupt financial intermediation when financial institutions raise capital from markets using a taxonomy different from the markets in which they invest. This is particularly the case when capital is raised in advanced economies with more ambitious taxonomies than the emerging markets in which the capital is to be invested.

Underdeveloped financial markets constrain access to capital for climate-aligned projects. Lack of access to adequate local financing is particularly pronounced in some developing Asian economies where capital markets are less developed. The benefits of a developed capital market include a broader investor base and greater transparency to facilitate risk-sharing and mobilize large amounts of capital via direct financing (IMF 2023; NGFS 2022). Some climate adaptation projects are infrastructure investments in nature and will benefit from long-term local currency financing to reduce currency and maturity mismatches and related risks. Financiers would thus benefit from a functioning local currency capital market. Meanwhile, international institutional investors face difficulties in identifying relatively liquid assets in the emerging-market and developing economies that meet their climate mandates. As functioning capital markets also provide more liquid assets, lack of functioning capital markets limits private capital to only emerging-market and developing economies with large and relatively liquid capital markets.

4.4.2.2 Developing Climate-Enabling Financial Systems

"Greening" financial systems can encourage accurate pricing of climate risks and attract capital for climate-aligned projects. A climate-aligned financial system promotes private climate capital by mainstreaming systemic climate-risk pricing in investments. Policies to green the financial sector help mitigate climate-related risks to safeguard macrofinancial stability and encourage investments in low-carbon activities (Gasparini and Tufano 2023; Carrasco and Lee 2023; Shirai 2023). The integration of climate risks into macroprudential considerations provides regulatory incentives for investors to price climate-related risks when evaluating project investability, making climate-aligned projects more attractive (Brunnermeier and Landau 2020). Global central banks have therefore introduced various financial, monetary, and additional policies to green the financial system. Table 4.2 lists such monetary and finance policies implemented by 32 global central banks. European central banks and financial supervisors have implemented the world's most comprehensive green financial and monetary policy tools (Gupta, Cheng, and Rajan 2023), followed by Asian central banks, then select Latin American economies (Dikau and Volz 2018; Durrani, Rosmin, and Volz 2020).

Table 4.2: "Green" Monetary and Financial Policies of Global Central Banks, 2021

FINANCIAL POLICY	**Credit Guidance Policy**	Priority sector lending policy
	Prudential Policy	Preferential green capital requirements
	Supervisory Policy	Climate stress test
		Disclosure requirements
		Supervisory expectations and guidelines
		Survey of practices
MONETARY POLICY	**Collateral**	Incorporating climate-related criteria for accepting collateral
	Credit operations	Encouraging green finance
	Domestic asset purchases	Incorporating climate criteria in own asset management and financial portfolio of central banks
	Foreign asset purchases	Incorporating climate criteria in own foreign exchange reserve management
ADDITIONAL POLICIES	**Change in mandate**	Explicit change in de jure mandate of the central bank to include climate change or transition of the economy to net zero
	Other policies	Miscellaneous (green bond guidelines, inclusion of climate-related financial risks in annual report of central bank, inclusion of ESG risks by financial sector supervisors)
	Own funds	Own asset and portfolio management by central banks

ESG = environmental, social, and governance.
Source: E-Axes Forum. Green Monetary and Financial Policies (GMFP) Tracker (accessed 30 May 2024).

A "greener" financial sector, combined with deeper financial sector development, enables more private climate capital. According to Buchner et al. (2023), domestic corporations and financial institutions (both private and state-owned) provided 91.9% of total climate finance flows in East Asia and the Pacific for 2021 and 2022. Around 96% of these total flows were financed from domestic sources. This is much higher than the corresponding levels in South Asia during the same period, where 51.4% of total climate finance was from domestic corporations and financial institutions, of which about 43% was financed from domestic funding sources. The higher share of climate finance from corporations and financial institutions, as well as from domestic funding sources, in East Asia is related to the subregion's more-comprehensive green finance policies (Figure 4.16). Deeper financial sector development in East Asia has also enabled green finance policies to supplement climate policies in channeling private climate capital.

Figure 4.16: "Green" Policies and Regulations in Selected Developing Asian Economies

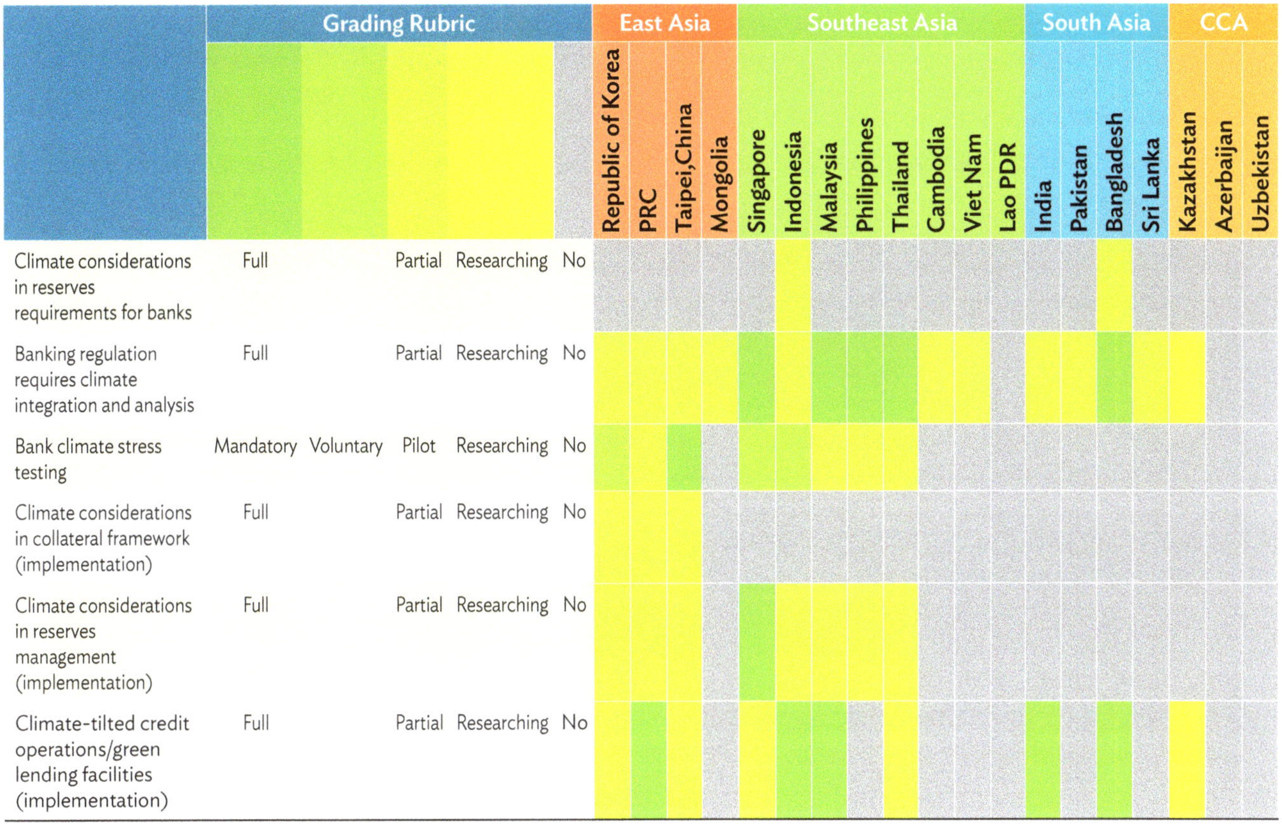

The figure presents a color-coded grading matrix across six policy/regulation dimensions (rows) and selected economies grouped by region (columns). Column groups and headers are:

- **Grading Rubric**
- **East Asia:** Republic of Korea, PRC, Taipei,China, Mongolia
- **Southeast Asia:** Singapore, Indonesia, Malaysia, Philippines, Thailand, Cambodia, Viet Nam, Lao PDR
- **South Asia:** India, Pakistan, Bangladesh, Sri Lanka
- **CCA:** Kazakhstan, Azerbaijan, Uzbekistan

Rows and their grading rubric scales:

Policy/Regulation	Grading rubric
Climate considerations in reserves requirements for banks	Full — Partial — Researching — No
Banking regulation requires climate integration and analysis	Full — Partial — Researching — No
Bank climate stress testing	Mandatory — Voluntary — Pilot — Researching — No
Climate considerations in collateral framework (implementation)	Full — Partial — Researching — No
Climate considerations in reserves management (implementation)	Full — Partial — Researching — No
Climate-tilted credit operations/green lending facilities (implementation)	Full — Partial — Researching — No

CCA = Caucasus and Central Asia, PRC = People's Republic of China, Lao PDR = Lao People's Democratic Republic.
Source: Mortimer, J. and S. Tian. 2024. Mobilizing Private Climate Finance – The Role of Climate Policies. Background paper for the *Asia–Pacific Climate Report 2024: Catalyzing Finance and Policy Solutions.* Asian Development Bank.

Accelerating climate disclosure practice will help strengthen market integrity and attract private climate capital. Comparable and material climate disclosure leads to more-efficient pricing of climate-related risks, reduces transaction and compliance costs, and boosts investor confidence by lowering reputational concerns over greenwashing. Strengthening climate disclosure is an important agenda for developing Asia. While economies and institutions worldwide are still exploring the adoption of climate disclosure practices, the region would benefit from accelerating the adoption of globally compatible disclosure practices, including the ISSB standards, while upholding their integrity as a globally compatible baseline for disclosures. Timely adoption of climate disclosure helps regional companies to become competitive in attracting global sustainable capital and increasing their market share because investors and clients increasingly require comparable climate sustainability data and transition plans (Mortimer and Tian 2024).

Governments need to equip companies with the necessary resources and capacity to accelerate climate disclosure. ISSB standards have been developed as a globally compatible baseline for businesses of all sizes, sectors, and regions, but micro, small, and medium-sized enterprises and developing economies face resource constraints when beginning to implement these standards. An overly demanding climate regulatory regime may lead to compliance-averse attitudes and regulatory paralysis among market participants. Policymakers must therefore balance the market integrity gained from accelerating climate disclosure with the regulatory burden imposed on firms. To address this, the ISSB provides transition standard relief to all entities in the first year of reporting for certain requirements. Various capacity-building initiatives and resources have been developed for supporting both regulators and micro, small, and medium-sized enterprises to adopt and meet these climate reporting requirements (IFRS 2024; ACCA 2024). Implementing climate disclosure, together with market supervision from institutional investors, can boost market integrity via enhanced market self-regulation.

Governments should address challenges around transition finance to channel more private capital to relevant activities and projects. As indicated by ICMA (2024) and EY Global Financial Services (2024), it is important that regional regulators develop aligned or generally agreed definitions of transition finance and credible transition taxonomies. This can potentially unlock more cross-border financing while expanding the overall market for transition finance. Unlike green or zero emissions investments, there is no commonly accepted definition and criteria for differentiating legitimate transition assets from business-as-usual assets. It means regulators should support the development of detailed technical standards, either in the form of sectoral technology road maps or transition taxonomies with technical and temporal screening thresholds, to mitigate the reputational risk concerns of investors, who may fear that investing in hard-to-abate transition projects will lead to greenwashing accusations (Mortimer and Tian 2024). Governments should guide the corporate transition plan frameworks and related disclosures so that corporate entities can communicate with investors to show their transition strategies are feasible and credible based on officially recognized road maps. To address the lack of economic viability in coal phaseout projects and the shortage of successful decarbonization demonstrations, governments can use climate policies and public finance resources to generate incentives, make structural changes in the markets, and de-risk related projects at a sufficient level, thus fostering greater private sector participation and innovative solutions for decarbonization.

Harmonized taxonomies can facilitate international climate investment. Harmonized sustainable finance taxonomies could unlock a global private finance pool by reducing transaction costs for investors. One example is the Common Ground Taxonomy (CGT) of the International Platform on Sustainable Finance, which maps and compares the commonalities between the European Union Taxonomy and the PRC Catalogue. This analysis makes a significant initial step toward aligning global standards for green and sustainability definitions. Although the CGT is not legally binding for the European Union or the PRC, it serves as a valuable tool for climate financing and investments. By providing a means to compare criteria and thresholds of the two taxonomies for various economic activities, it helps establish interoperability and promote investment flows.

4.4.3 De-Risking Climate Projects Using Public Finance

4.4.3.1 Barriers to Private Climate Investment

Some projects face sector-specific challenges of attracting private climate investment. Some sectors benefit from standardized documents that have been repeatedly used around the world on similar commercial terms, making it easier for investors to evaluate projects. For example, renewable energy projects benefit from standardized documents such as power purchase agreements. In addition, loan documentation follows the templates issued by the Loan Market Association or the Loan Syndications & Trading Association. Such agreements enable private investors to finance different projects with similar contractual documentation. Standardization is not prevalent in other sectors, such as flood control and irrigation, which have seen limited private sector involvement to date (Mik 2024). Meanwhile, some sectors such as renewable energy generation and electric vehicle manufacture rely on a few critical minerals, namely lithium, copper, cobalt, and rare earths (Leruth et al. 2022; Hund et al. 2020).[12] Hund et al. (2020) estimate that the net zero transition could lead to a 500% increase in the production of critical minerals. This brings supply challenges due to insufficient mining capacity. Huge investment costs and lead times of 10–15 years will be needed before the supply of critical minerals can meet demand (Manalo 2023). Moreover, supplies of critical minerals have been found to be concentrated in only a few economies (Figure 4.17). Such concentration has exacerbated supply constraints.

---

[12] For instance, solar panels and wind turbines use aluminum, copper, silver, silicon, and rare earth metals. Electric vehicles running on rechargeable batteries require lithium, manganese, nickel, and cobalt for their production. Rare earths are vital in the manufacture of high-tech semiconductors, batteries, and renewable energy (Leruth et al. 2022).

Figure 4.17: Share of the Top Three Economies in Total Production of Selected Minerals, 2022

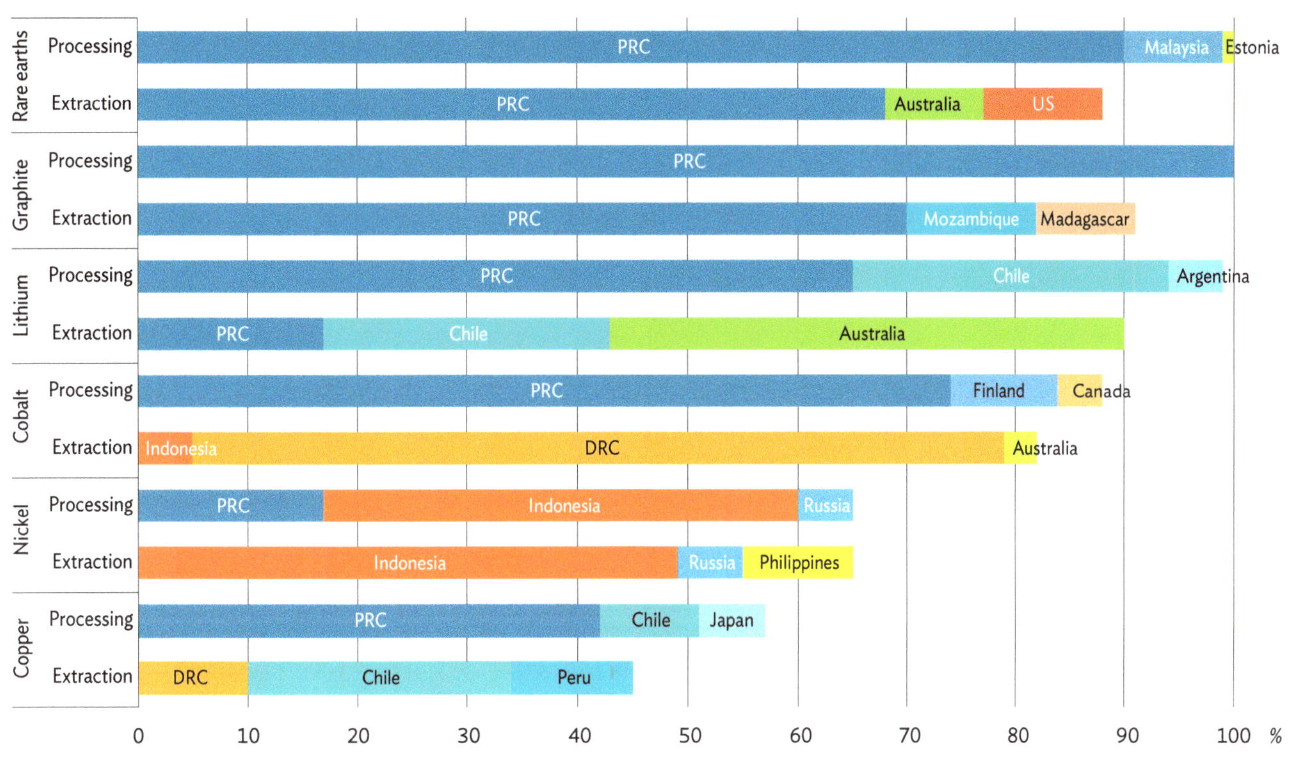

PRC = People's Republic of China, DRC = Democratic Republic of the Congo, US = United States.
Note: Graphite extraction is for natural flake graphite. Graphite processing is for spherical graphite for battery grade.
Source: International Energy Agency. Share of Top Three Producing Countries in Mining of Selected Minerals, 2022 (accessed 21 June 2024).

Some climate projects are commercially unviable due to lack of insurance, high upfront costs, and long payback horizons. Some climate-related projects, due to their very nature, are definitively uninsurable: other projects may not be insurable in practical terms due to exorbitant premiums and deductibles (Mik 2024). Repeated losses due to climate-related risk exposure may also lead to the cancellation of insurance. Many climate investments, especially those that qualify as infrastructure, require sizable investments and time-consuming feasibility work and preparations to receive permits. For example, upfront capital and financing costs can account for 70%–80% of the levelized cost of electricity generation for wind projects and 80%–90% for solar photovoltaic systems (IEA 2020). Many climate projects have long payback periods, sometimes measured in decades, which do not compare favorably to other investment opportunities, such as listed equities and bonds.

The commercial viability of climate investments can also be weakened by high preparation costs, socioenvironmental costs, and market risks. Climate-related projects can involve high transaction and preparation costs, including upstream investments in transmission capacity, construction of new infrastructure, and land acquisition, among others (Mik 2024). Private developers face opposition from local communities upset about socioenvironmental impacts, such as disruptions to traffic, changing visual impacts, and noise and dust (Garcia 2020). Private investors also have to charge risk premiums to cover political risk, regulatory risk, currency risk, and transfer risk when investing in emerging economies (IEA 2023). Climate-oriented infrastructure projects face additional risks beyond the conventional risk factors such as politics, currency, interest rates, land acquisition, and labor relations (IEA and IFC 2023).[13] Many climate-oriented projects produce "public goods", which do not have a proper business model for private investors because the many beneficiaries do not pay for services such as flood control (Mik 2024).

---

13   See also International Energy Agency. World Energy Investment 2023.

Additional market rigidities prevent private investments in climate-oriented projects. ADB (2023c) highlights a few market rigidities that disincentivize private investment in climate-oriented projects across developing Asia. Because the electricity sector in many Asian economies is regulated, tariffs may not be set based on underlying production costs and this may impact investors' returns. This scenario weakens the price correction of carbon pricing as an incentive to make low-carbon investments. Another typical barrier is the contractual obligation in tenders that commits to using emissions-intensive technologies. For example, long-term power purchase agreements between utilities and power producers shield existing coal-fired power plants from market risks with a guaranteed revenue. Investors are confident that they will not become stranded assets (Trencher et al. 2020). Such long-term obligations reduce the demand for low-carbon projects and disincentivize investments in decarbonizing activities.

4.4.3.2 Using Public Resources to Address Barriers

The public sector can increase the viability of climate-oriented projects with intervention such as structural changes to markets. For example, to improve the commercial viability of renewable energy projects and incentivize private investment, the public sector has enacted various interventions, such as purchase mandates, preferential tariffs, super priority in the merit order, unbundling utilities, socialized costs of connection, and net metering, among other measures. These measures have had a detrimental impact on the financial performance of vertically integrated public utilities (Mik 2024). The public sector can strengthen the investment environment to enable private sector participation, by addressing entry barriers and early-stage challenges to help scale climate-related investment, especially in new technology areas. The public sector can also work to overcome market rigidities such as long-term power purchase agreements that delay decarbonizing investments (Iyer and Vaze 2024).

Various de-risking instruments can improve the commercial viability of climate projects. Governments, development banks, and financial institutions can use innovative de-risking financing solutions to alter cash flow patterns of climate-related projects, thus encouraging different types of private sector investors. One commonly used solution is blended finance, which is the practice of blending public finance and philanthropic funding with private capital to reduce the risk exposure of private investors. In practice, governments provide funding in the form of grants, technical assistance, or concessional financing (debt and equity) to de-risk the project when investors cannot get their returns. Typical de-risking financial instruments include first-loss guarantees, minimum-revenue guarantees, credit enhancements or guarantees, junior or subordinated concessional capital, hedging, quasi-insurance, and partial risk guarantees, among others (Iyer and Vaze 2024; Mik 2024; CISL 2023). These instruments reduce private investors' risk exposure and attract investors with different risk-return preferences and mandates, such as traditional and risk-averse private investors (CSIS 2023). In doing so, blended finance can leverage more private capital and diversify funding sources for climate investments. Some blended finance schemes also offer nonfinancial levers to help address barriers to private climate investment, via knowledge sharing and capacity building.

4.5   Global Initiatives and the Role of Multilateral Organizations

Initiatives set at the global level provide levers to mobilize private climate finance. The first global stocktake of the Paris Agreement was conducted by the United Nations Framework Convention on Climate Change in 2023. This vital review highlighted an emissions gap—indicating that current climate commitments are not aligned with the 1.5°C goal—and underscored the need for system-wide transformations (WRI 2023; UNFCCC 2023). Nationally determined contributions (NDCs) are slated to be updated in 2025, with a call to ramp up commitments, and the second global stocktake will take place in 2028. Meanwhile, initiatives such as the determination of a New Collective Quantified Goal on Climate Finance, the Task Force for the Global Mobilization Against Climate Change, an initiative under the Brazilian Presidency of the G20, the NDC Partnership, and the Coalition for High Ambition Multilevel Partnerships for Climate Action all recognize the need to engage with the private sector to scale up finance and create innovative financing mechanisms.

Voluntary networks are inspiring greater climate commitments from financial institutions and investors. Global climate networks and coalitions provide a range of support to their members, including capacity building on net zero policies and regulations, sharing best practices, and setting collective investment targets. The Glasgow Financial Alliance for Net Zero is a global coalition of 675 leading financial institutions committed to an accelerated low-carbon transition. The Network for Greening the Financial System, which started as a collaborative platform for eight central banks, has now grown to 138 members and aims to enhance the role of the financial system in managing risks and to mobilize capital for green and low-carbon investments. The Principles for Responsible Investment is an investor-led initiative—with support from the United Nations Environment Programme's Finance Initiative and the United Nations Global Compact— whose 7,000 signatories across 135 economies have committed to fostering good governance and promoting investments aligned with environmental sustainability.

Multilateral organizations play a catalytic role in mobilizing private finance. The Organisation for Economic Co-operation and Development (OECD 2023b) reports a $13.54 billion annual average of private finance leveraged globally from multilateral and bilateral blended finance transactions made across 2016–2020. Of this amount, 39% or $5.2 billion in private capital was leveraged in Asia, using various mechanisms such as guarantees, syndicated loans, credit lines, cofinancing, and shares and investments in special or collective financing vehicles. Collectively and individually, multilateral development banks (MDBs) are serving as catalytic partners in private sector capital mobilization for climate-oriented projects (Viewpoint Note 2024). They are also using other modalities, such as policy-based lending, to enable reforms that address policy bottlenecks in private investment. MDB facilities, such as ADB's Asia Pacific Project Preparation Facility, provide technical support and financial resources to strengthen the bankability of projects in developing economies. MDBs are additionally helping create enabling environments for private investments and build institutional capacities for investment pipeline development. Through technical assistance and advisory services, MDBs are promoting project pipelines and modalities suited to private investor appetites and are identifying partnerships with global initiatives to facilitate climate-oriented projects. Catalytic MDB and public sector support for raising private climate capital is particularly relevant in economies with less-developed capital markets. Other economies with deeper and broader capital markets leave more room for private sector engagement.

References

ACCA. 2024. Sustainability Reporting – SME Guide 6th Final.

ADB (Asian Development Bank). 2020. *Climate Change, Coming Soon to a Court Near You. Climate Litigation in Asia and the Pacific and Beyond.*

_____. 2022. *Survey on Green Bonds and Sustainable Finance in ASEAN: Insights on the Perspectives of Institutional Investors and Underwriters.*

_____. 2023a. *Climate Finance Landscape of Asia and the Pacific.*

_____. 2023b. Climate Change Financing at ADB, 2023.

_____. 2023c. Update on ADB's Energy Transition Mechanism - April 2023.

_____. 2023d. *Asia in the Global Transition to Net Zero: Asian Development Outlook 2023 Thematic Report.*

_____. 2024. *Asian Economic Integration Report 2024: Decarbonizing Global Value Chains.*

Akomea-Frimpong, I., D. Adeabah, D. Ofosu, and E. J. Tenakwah. 2022. A Review of Studies on Green Finance of Banks, Research Gaps and Future Directions. *Journal of Sustainable Finance & Investment.* 12 (4). pp. 1241–1264.

Alam, A., A. M. Du, M. Rahman, H. Yazdifar, and K. Abbasi. 2023. SMEs Respond to Climate Change: Evidence from Developing Countries. *Technological Forecasting and Social Change.* 185 (2022). 122087.

Ando , S., F. Chen, F. Roch, and U. Wiriadinata. 2023. How Large Is the Sovereign Greenium? *International Monetary Fund Working Paper.* WP/23/80.

Arca and SGX RegCo. 2024. Consultation Paper on the Recommendations by the Sustainability Reporting Advisory Committee.

Ardia, D., K. Bluteau, K. Boudt, and K. Inghelbrecht. 2021. Climate Change Concerns and the Performance of Green Versus Brown Stocks. National Bank of Belgium Working Paper.

Atz, U., T. Van Holt, Z. Z. Liu, and C. Bruno. 2022. Does Sustainability Generate Better Financial Performance? Review, Meta-analysis, and Propositions. *Journal of Sustainable Finance and Investment.*

Bangladesh Bank. 2023. *Guideline On Sustainability and Climate-Related Financial Disclosure.*

BCG (Boston Consulting Group). 2023. What Investors' Attitudes Reveal About the Future of Climate Finance.

BEIS (Government of the United Kingdom, Department for Business, Energy & Industrial Strategy). 2022. *Mandatory Climate-Related Financial Disclosures by Publicly Quoted Companies, Large Private Companies and LLPs.*

Berrou, R., P. Dessertine, and M. Migliorelli, M. 2019. An Overview of Green Finance. In M. Migliorelli, and P. Dessertine, eds. The Rise of Green Finance in Europe. Palgrave Studies in Impact Finance. Palgrave Macmillan.

Beyene W., K. de Greiff, M. Delis, and S. Ongena. 2021. Too-Big-To-Strand? Bond Versus Bank Financing in the Transition to a Low-Carbon Economy. Centre for Economic Policy Research Discussion Paper. No. DP16692.

Bloom, N. 2009. The Impact of Uncertainty Shocks. *Econometrica*. 77 (3). pp. 623–685.

Bloomberg. 2024. China Proposes New ESG Rules to Keep Up with Europe.

Bloomberg NEF. 2024. Energy Transition Investment Trends 2024.

Bowman, M. 2022. Turning Promises into Action: Legal Readiness for Climate Finance and Implementing the Paris Agreement. *Carbon & Climate Law Review*. 16 (1). pp. 41–55.

Brunnermeier, M. and J. P. Landau. 2020. Central Banks and Climate Change. Centre for Economic Policy Research. 15 January.

Buchner, B., B. Naran, R. Padmanabhi, S. Stout, C. Strinati, D. Wignarajah, G. Miao, J. Connolly, and N. Marini. 2023. Global Landscape of Climate Finance. Climate Policy Initiative. 2 November.

Calster, G., W. Vandenberghe, and L. Reins. 2022. *Research Handbook on Climate Change Mitigation Law*.

Carrasco, B. and J. Lee. 2023. Greening the Financial System: Climate Financial Risks and How ADB Can Help. ADB.

Cheng, B., I. Ioannou, and G. Serafeim. 2014. Corporate Social Responsibility and Access to Finance. *Strategic Management Journal*. 35 (1). pp 1–23.

Chidambaram, R. and P. Khanna. 2022. It's Time to Invest in Climate Adaptation. Harvard Business Review. 1 August.

China Council for International Cooperation on Environment and Development. 2022. Green Consensus and High Quality Development.

Choi, D., Z. Gao, and W. Jiang. 2020. Attention to Global Warming. *Review of Financial Studies*. 33. pp. 1112–1145.

Ciminelli et al. Forthcoming. Climate Law and Sustainable Fund Holdings.

CISL (University of Cambridge Institute for Sustainability Leadership). 2023. Everything, Everywhere, All at Once: How Can Private Finance be Unlocked for Nature and Climate in the International Financial Architecture?

Climate Governance Initiative. 2023. Navigating the Climate Disclosure Landscape.

Climate Policy Initiative. 2023. Global Landscape of Climate Finance 2023.

Cox, E., C. Kelly, B. Murhpy, and N. Röttmer. 2022. Time to Get Serious About the Realities of Climate Risk. PwC.

Crumpler, K., R. Abi Khalil, E. Tanganelli, N. Rai, L. Roffredi, A. Meybeck, V. Umulisa, J. Wolf, and M. Bernoux. 2021. 2021 (Interim) Global Update Report – Agriculture, Forestry and Fisheries in the Nationally Determined Contributions. *Environment and Natural Resources Management Working Paper*. No. 91. Food and Agriculture Organization of the United Nations.

CSIS (Centre for Strategic and International Studies). 2023. Private Capital Mobilization for Climate Finance in an International Context.

Dahlqvist, F., S. Kane, L. Leinert, M. Moosburger, and A. Rasmussen. 2023. Climate Investing: Continuing Breakout Growth Through Uncertain Times. McKinsey & Company. 13 March.

Deloitte. 2024. Gen Z and Millennial Survey: Living and Working with Purpose in a Transforming World.

Diaz-Rainey, I., P. A. Griffin, D. H. Lont, A. J. Mateo-Márquez, and C. Zamora-Ramírez. 2023. Shareholder Activism on Climate Change: Evolution, Determinants, and Consequences. *Journal of Business Ethics*. 193. pp. 481–510.

Dikau, S. and U. Volz. 2018. Central Banking, Climate Change and Green Finance.

Du, J., H. Wenxiu, and W. Zhang. 2020. Corporate Social Responsibility Information Disclosure and Innovation Sustainability: Evidence from China. *Sustainability*. 12 (1). p. 409.

Durrani, A., M. Rosmin, and U. Volz. 2020. The Role of Central Banks in Scaling Up Sustainable Finance - What Do Monetary Authorities in the Asia-Pacific Region Think? *Journal of Sustainable Finance & Investment*. 10 (2). pp. 92–112.

Eccles, J. 2021. Moving Beyond Modern Portfolio Theory: It's About Time! *Forbes*. 18 May.

The Economist. 2020. How Much Can Financiers Do About Climate Change? 20 June.

The Economist. 2021. The Private Sector Starts to Invest in Climate Adaptation. 22 July.

Ehlers, T., C. Gardes-Landolfini, F. Natalucci, and P. Ananthakrishnan. 2022. How to Scale Up Private Climate Finance in Emerging Economies. *Global Financial Stability Report*. International Monetary Fund.

Ehlers, T., F. Packer, and K. de Greiff. 2021. The Pricing of Carbon Risk in Syndicated Loans: Which Risks are Priced and Why? *Journal of Banking and Finance*. 136. March.

Engle, R. F., S. Giglio, B. T. Kelly, H. Lee, and J. Stroebel. 2020. Hedging Climate Change News. *Review of Financial Studies*. 33. pp. 1184–1216.

ERB (External Reporting Board). 2022. Climate-Related Disclosures. External Reporting Board.

ERM (Environmental Resources Management) and Ceres. 2021. The Changing Climate for Private Equity. Sustainability Report 2021.

EU (European Union). 2022. Directive (EU) 2022/2464 of the European Parliament and of the Council. *Official Journal of the European Union*.

EY Global Financial Services. 2024. How to Accelerate Transition Finance for Net Zero.

Flynn, H., D. Morales, J. Delongchamp, and C. Hugger. 2023. Why Climate Change Matters in Private Markets. Wellington Management.

Garcia, R. 2020. L'opposition aux Projets éoliens Terrestres, une Forme de Contestation «territorialisée». *Géographie et Cultures*. 114. pp. 135–152.

Gasparini, M. and P. Tufano. 2023. The Evolving Academic Field of Climate Finance.

Gavriilidis, K. 2021. *Measuring Climate Policy Uncertainty*.

GRI (Global Reporting Initiative). 2022. Climate Reporting in ASEAN: State of Corporate Practices.

Gupta, B., R. Cheng, and R. S. Rajan. 2023. Green Financial and Regulatory Policies: Why Are Some Central Banks Moving Faster than Others? *Global Environmental Politics*. 23 (4). pp. 73–93.

He, G. 2023. Asian Climate Policies and Way Forward. ADB.

Heinkel, R., A. Kraus, and J. Zechner. 2001. The Effect of Green Investment on Corporate Behavior. *Journal of Financial and Quantitative Analysis*. 36. (4). pp. 431–449.

Henderson, D., A. Ghosh, and J. Zhan. 2018. Corporate Environmental Responsibility and Firm Risk. *Academy of Management Journal*. International Energy Agency.

Hengge, M., U. Panizza, and R. Varghese. 2023. Carbon Policy Surprises and Stock Returns: Signals from Financial Markets. SSRN. 4343984.

HKEX (Hong Kong Stock Exchange). 2023. Enhancement of Climate-Related Disclosures Under the Environmental, Social and Governance Framework. Hong Kong Stock Exchange.

Hund, K., D. La Porta, T. P. Fabregas, T. Laing, and J. Drexhage. 2020. Minerals for Climate Action: The Mineral Intensity of the Clean Energy Transition. World Bank.

Hyun, S., D. Park, and S. Tian. 2021. Pricing of Green Labeling: A Comparison of Labeled and Unlabeled Green Bonds. *Finance Research Letters*. 41. 101816.

ICMA (International Capital Market Association). 2024. *Transition Finance in the Debt Capital Market*.

IEA (International Energy Agency). 2017. *Creating Markets for Climate Business: An IFC Climate Investment Opportunities Report*.

_____. 2020. *Energy Investing: Exploring Risk and Return in the Capital Markets.* A Joint Report by the International Energy Agency and the Centre for Climate Finance & Investment.

_____. 2022. Global EV Outlook 2022.

_____. 2023. Cost of Capital Survey Shows Investments in Solar PV Can Be Less Risky than Gas Power in Emerging and Developing Economies, Though Values Remain High. International Energy Agency. 30 November.

IEA (International Energy Agency) and IFC (International Finance Corporation). 2023. *Scaling Up Private Finance for Clean Energy in Emerging and Developing Economies*. International Energy Agency.

IFC (International Finance Corporation). 2017. *Climate Investment Opportunities in South Asia: An IFC Analysis*.

IFRS. 2024. Jurisdictional Sustainability Consultations (accessed 29 August 2024).

IMF (International Monetary Fund). 2019. Sustainable Finance: Looking Farther (Chapter 6) in *Global Financial Stability Report: Lower for Longer*.

______. 2020. Global Financial Stability Report.

______. 2023. *Global Financial Stability Report: Financial and Climate Policies for a High-Interest-Rate Era.*

ISSB (International Sustainability Standards Board). 2023. IFRS S2 IFRS® Sustainability Disclosure Standard: Climate-Related Disclosures.

Iyer, K. and P. Vaze. 2024. Landscape of Blended Finance and Role of Public Finance. Background paper for the *Asia–Pacific Climate Report 2024: Catalyzing Finance and Policy Solutions.* ADB.

Kagda, S. 2024. Transition Finance Critical for South-East Asia to Achieve Net-Zero Goals: Panel. *The Business Times.* 18 March.

Kawabata, T. 2019. What are the Determinants for Financial Institutions to Mobilise Climate Finance? *Journal of Sustainable Finance & Investment.* 9 (4). pp. 263–281.

Koerner, S. A., W. S. Siew, A. A. Salema, P. Balan, S. Mekhilef, and N. Thavamoney. 2022. Energy Policies Shaping the Solar Photovoltaics Business Models in Malaysia with Some Insights on Covid-19 Pandemic Effect. *Energy Policy.* 164 (2022). 112918.

Laidlaw, J. 2024. June 2024 – Where Does the World Stand on ISSB Adoption? S&P Global. 19 July.

Leruth, L., P. Regibeau, L. Renneboog, and A. Mazarei. 2022. *Green Energy Depends on Critical Minerals. Who Controls the Supply Chains?* Peterson Institute for International Economics.

Lim, C. H., R. Basu, Y. Carrière-Swallow, K. Kashiwase, M. Kutlukaya, M. Li, E. Refayet, D. Seneviratne, M. Sy, and R. Yang. 2024. Unlocking Climate Finance in Asia-Pacific. IMF Departmental Paper.

Majersik, C. 2019. Nareit's Leader in the Light Awards Show Innovative REITs Stepping Up on Climate Action. Institute for Market Transformation. 19 November.

Manalo, P. 2023. Discovery to Production Averages 15.7 years for 127 Mines. S&P Global. 6 June.

Mik, J. 2024. Strengthening Policies on Climate in Asia and the Pacific Through Economic Research, 2023-2025: Impediments to Private Climate Finance. Background paper for the *Asia–Pacific Climate Report 2024: Catalyzing Finance and Policy Solutions.* ADB.

J. P. Morgan. 2023. Asia Pacific Equity Research. 23 January.

Morita, T. and C. Pak. 2018. Legal Readiness to Attract Climate Finance: Towards a Low-Carbon Asia and the Pacific. CCLR. 12. p. 6.

Mortimer, J. and S. Tian. 2024. Mobilizing Private Climate Finance – The Role of Climate Policies. Background paper for the *Asia–Pacific Climate Report 2024: Catalyzing Finance and Policy Solutions.* ADB.

MSISI (Morgan Stanley Institute for Sustainable Investing). 2024. *Sustainable Reality.*

Nascimento, L. and N. Höhne. 2023. Expanding Climate Policy Adoption Improves National Mitigation Efforts. *npj Climate Action.* 2 (12).

NGFS (Central Banks and Supervisors Network for Greening the Financial System). 2022. Enhancing Market Transparency in Green and Transition Finance. Network for Greening the Financial System.

————. 2023. *Climate-Related Litigation: Recent Trends and Developments.* Network for Greening the Financial System.

OECD (Organisation for Economic Co-operation and Development). 2023a. *Private Finance Mobilised by Official Development Finance Interventions.* Development Co-operation Directorate, OECD Publishing.

————. 2023b. *Scaling Up the Mobilization of Private Finance for Climate Action in Developing Countries.*

————. 2023c. *Annual Survey of Investment Regulation of Pension Providers.*

Olasehinde-Williams, G., O. Özkan, and S. S. Akadiri. 2023. Effects of Climate Policy Uncertainty on Sustainable Investment: A Dynamic Analysis for the US. *Environmental Science and Pollution Research.* 30 (19). pp. 55326–55339.

ORF (Observer Research Foundation). 2023. *Bridging the Climate Finance Gap: Catalysing Private Capital for Developing and Emerging Economies.* Observer Research Foundation.

Pástor, L., R. F. Stambaugh, and L. A. Taylor. 2021. Dissecting Green Returns. National Bureau of Economic Research.

PWC. 2023. State of Sustainability Reporting in Asia. National University of Singapore, Centre for Governance and Sustainability.

Randall, T., J. Sedemund, and W. Bartz-Zuccada. 2023. Private Investment for Climate Change Adaptation – Difficult to Finance or Difficult to See the finance? London School of Economics and Political Science. 16 March.

Robins, N., V. Brunsting, and D. Wood. 2018. *Climate Change and the Just Transition: A Guide for Investor Action.* Grantham Research Institute on Climate Change and the Environment.

Roslan, F., Ş. C. Gherghina, J. Saputra, M. N. Mata, F. D. Mohmad Zali, and J. Moleiro Martins. 2022. A Panel Data Approach Towards the Effectiveness of Energy Policies in Fostering the Implementation of Solar Photovoltaic Technology: Empirical Evidence for Asia-Pacific. *Energies.* 15 (10). 3775.

Royal Government of Bhutan. 2021. Kingdom of Bhutan Second Nationally Determined Contribution.

Setiawan, A. D., T. N. Zahari, F. J. Purba, A. O. Moeis, and A. Hidayatno. 2022. Investigating Policies on Increasing the Adoption of Electric Vehicles in Indonesia. *Journal of Cleaner Production.* 380. 135097.

Setzer, J. and C. Higham. 2023. *Global Trends in Climate Change Litigation: 2023 Snapshot.* Grantham Research Institute on Climate Change and the Environment; and Centre for Climate Change Economics and Policy, London School of Economics and Political Science.

Sharfman, M. P. and C. S. Fernando. 2008. Environmental Risk Management and the Cost of Capital. *Journal of Strategic Management.* 29. pp. 569–592.

Shirai, S. 2023. *Green Central Banking and Regulation to Foster Sustainable Finance.* No. 1361. ADB.

Songwe V., N. Stern, and A. Bhattacharya. 2022. *Finance for Climate Action: Scaling Up Investment for Climate and Development*. Grantham Research Institute on Climate Change and the Environment, London School of Economics and Political Science.

SustainAbility Institute 2021. Sustainability Report 2021.

Tall, A., S. Lynagh, C. Blanco Vecchi, P. Bardouille, F. Montoya Pino, E. Shabahat, V. Stenek, F. Stewart, S. Power, and C. Paladines. 2021. Enabling Private Investment in Climate Adaptation and Resilience: Current Status, Barriers to Investment and Blueprint for Action. World Bank.

Tang, V. 2023. Beyond Financing Gaps: Sizing the Decarbonization Investment Opportunity. Government of Singapore Investment Corporation.

Temple-West, P. and B. Masters. 2024. JP Morgan and State Street Quit Climate Group as BlackRock Scales Back. *Financial Times*. 16 February.

Treloar, S. 2023. Norway Approves More Than $18 Billion of Oil and Gas Projects. *Bloomberg*. 28 June.

Trencher, G., A. Rinscheid, M. Duygan, N. Truong, and J. Asuka. 2020. Revisiting Carbon Lock-In in Energy Systems: Explaining the Perpetuation of Coal Power in Japan. *Energy Research & Social Science*. 69. 101770.

UNDP (United Nations Development Programme). 2023. Launch of Fiji's First-Ever Sovereign Blue Bond.

UNFCCC (United Nations Framework Convention on Climate Change). Nationally Determined Contributions Registry (accessed 21 May 2024).

————. 2023. Yearbook of Global Climate Action 2023: Marrakech Partnership For Global Climate Action.

Viewpoint Note. 2024. MDBs Working as a System for Impact and Scale.

WEF (World Economic Forum). 2022. Climate Adaptation: The $2 Trillion Market the Private Sector Cannot Ignore. 1 November.

————. 2023. How to Finance the Energy Transition in Asia.

Weinert, J. 2023. Why Your Business Should Take Action? *Signify*. 29 September.

Wong, E. 2024. Channelling Sustainable Finance: The Role of Taxonomies. Background paper for the *Asia–Pacific Climate Report 2024: Catalyzing Finance and Policy Solutions*. ADB.

WRI (World Resources Institute). 2023. State of Climate Action 2023.

Wu, J., S. O'Shea, L. Kladitis, S. Abrahams, and R. Philips. 2023. Adapting to a Warmer Planet: Why Climate Change Investing Isn't Just About Decarbonisation. J. P. Morgan Asset Management.

Chapter 5

Carbon Pricing as a Key Policy for Climate Change Mitigation

5.1 Mounting Evidence in Support of Carbon Pricing

Carbon pricing is essential for achieving climate change mitigation in a cost-effective manner.
As discussed in Chapter 1, while developing Asia[1] has made significant progress, substantial decarbonization
remains a major challenge as the region continues to pursue economic growth. It is therefore crucial for
economies in the region to adopt emission reduction policies that minimize societal costs. By assigning
a price to greenhouse gases (GHGs), carbon pricing creates incentives for businesses to shift from
emissions-intensive products and processes to cleaner alternatives, spurring investment in low-carbon
technologies and encouraging consumers to opt for low-emission goods and services. It helps align market
behavior with climate objectives, making it a powerful and efficient mechanism for reducing GHG emissions
and driving the transition to a low-carbon economy.

**Robust evidence confirms that carbon pricing has been one of the most effective measures to reduce
GHG emissions.** A major study of 17 established carbon pricing policies found that such systems reduced
emissions by between 5% and 21% in their first few years of operation (Döbbeling-Hildebrandt et al. 2024).
Similar results were also found through empirical analysis across 143 economies, 43 of which had a carbon
price in place. The analysis showed that an additional price of €1 per metric ton of carbon dioxide (CO_2)
helped reduce subsequent annual emissions growth by approximately 0.3 percentage points (Best, Burke,
and Jotzo 2020). However, the performance of carbon pricing varies depending on the economy, sector,
and policy environment being assessed (IPCC 2022).

There is a broad range of carbon pricing instruments available to governments. Subsidies for producing
or consuming fossil fuels, as well as certain agricultural practices, can act as negative carbon pricing, which
further encourages GHG-emitting activities. There are, however, a variety of carbon pricing instruments that
discourage GHG emissions. The two primary instruments are carbon taxation and an emissions trading system
(ETS), with both being cost-effective policy instruments for achieving environmental goals. Both can provide
a clear carbon-price signal, either in terms of the cost (tax) of future emissions or the quantity of expected
emissions reductions. They can also generate additional revenue for governments, which can then be used
to support sustainable economic growth and "green" investment. International carbon markets, comprising
both compliance and voluntary markets, can also help reduce emissions cost-effectively by leveraging the
differences in abatement costs across economies.

Momentum is growing for the use of carbon pricing in the Asia and Pacific region.[2] Across the region,
eight national carbon pricing initiatives had been implemented or were pending implementation at the time
of publication. Japan and Singapore both employ carbon taxes, while Australia, the People's Republic of
China (PRC), Indonesia, Kazakhstan, the Republic of Korea (ROK), and New Zealand have all implemented
a national ETS. Subnational ETSs are also employed by local jurisdictions in Japan. The Philippines, Thailand,
and Viet Nam are considering adopting a domestic ETS, while India is developing a domestic national carbon
market. Momentum is also growing for economies in the region to take advantage of international carbon
markets, through both emerging compliance markets under Article 6 of the Paris Agreement on Climate
Change and the voluntary carbon market.

The choice of instruments, and their combination, depends on a jurisdiction's goals and circumstances.
This chapter discusses recent developments, best practices, and latest insights with regard to different carbon
pricing policies, taking into account the regional context. To some extent, the design and implementation of
carbon pricing policies are more important than the choice between the instruments.

---

[1]   In this chapter, developing Asia refers to the 46 developing members of ADB.
[2]   In this chapter, Asia and the Pacific refers to the 46 developing members of ADB, along with Australia, Japan, and New Zealand.

This chapter was written by a team of ADB staff consisting of Sanchita Basu-Das, Virender Kumar Duggal, Neil Foster-
McGregor, Yi Jiang (lead author), Jong Woo Kang, Kijin Kim, and Manisha Pradhananga.

5.2   Fossil Fuel Subsidies as Negative Carbon Pricing

5.2.1   High Public Financial Support for Fossil Fuels

In 2022, developing Asia provided $600 billion in public financial support for fossil fuels. Acting as a negative carbon price and reinforcing dependence on fossil fuels, this support included $440 billion in subsidies, at least $146 billion in capital investments by state-owned enterprises (SOEs), and $14 billion in international public finance (Figure 5.1). Together, this made up about a quarter of total global financial support to fossil fuels, which exceeded $1.9 trillion in 2022.[3] Public financial support for fossil fuels increases GHG emissions in two ways. First, subsidies for consumption directly reduce prices through measures such as price caps, direct transfers, and tax cuts, leading consumers to use more fossil fuels than they otherwise would. Second, support for new fossil fuel production capacity, for example by SOEs, incentivizes lock-in of new emissions-intensive infrastructure that must operate for many decades to recoup investments, creating "stranded asset" risks and slowing the transition to clean energy. A majority of fossil fuel subsidies in 2022 ($429 billion) were directed toward consumption, particularly that of electricity (43%)[4] and petroleum products (32%). Figure 5.2 shows that Kazakhstan, the Kyrgyz Republic, Tajikistan, Timor-Leste, and Uzbekistan dedicated a substantial share of national resources to fossil fuel subsidies in 2022.[5] Production subsidies amounted to around $11 billion, but this is likely to be an underestimate given many subsidies are undocumented or unqualified.

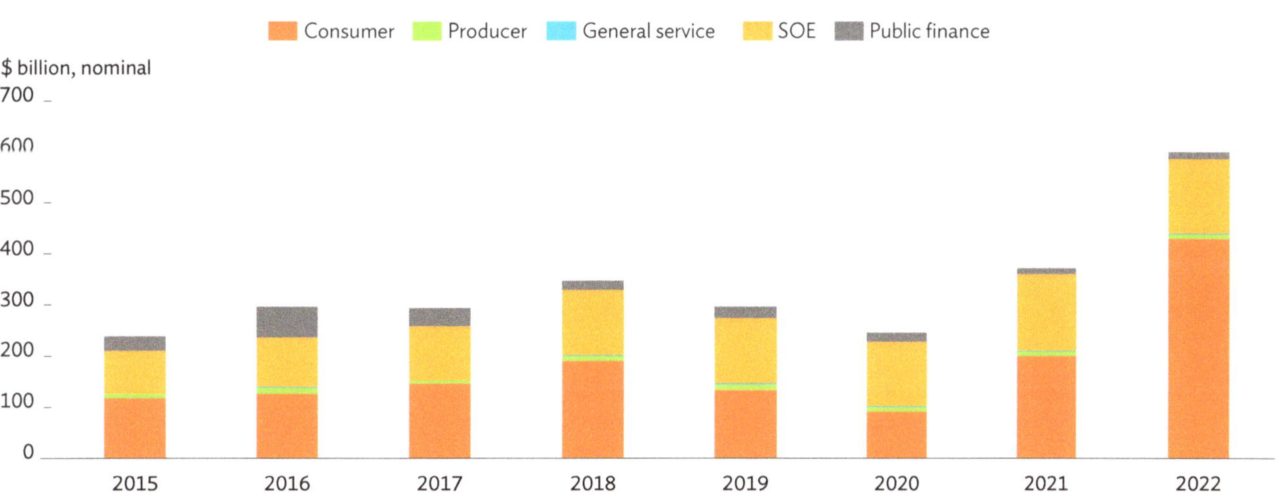

Figure 5.1: Public Financial Support for Fossil Fuels in Developing Asia, 2015–2022

SOE = state-owned enterprise.
Sources: International Institute for Sustainable Development and Organisation for Economic Co-operation and Development. 2024. Fossil Fuel Subsidy Tracker (accessed 29 May 2024); Laan, T., A. Geddes, N. Do, L. Cameron, S. Goel, and N. Jones. 2023. Burning Billions: Record Public Money for Fossil Fuels Impeding Climate Action. *Energy Policy Tracker*; and Oil Change International 2024.

---

3   Subsidies of $1.52 trillion (IISD and OECD 2024) plus capital investments by SOEs and lending from international public financial institutions, which totaled $0.35 trillion and $0.02 trillion, respectively, in G20 countries alone (Laan Geddes, Do, et al. 2023).
4   Electricity is considered a fossil fuel subsidy because the majority is generated from coal and natural gas.
5   Estimates of fossil fuel subsidies in this report include the use of the "price-gap" approach, which compares the end-use prices paid by fuel consumers with reference prices (such as import-parity prices). However, the ability to earn export or import-parity prices is often hindered by lack of sufficient export infrastructure, e.g., for natural gas and electricity from West and Central Asia.

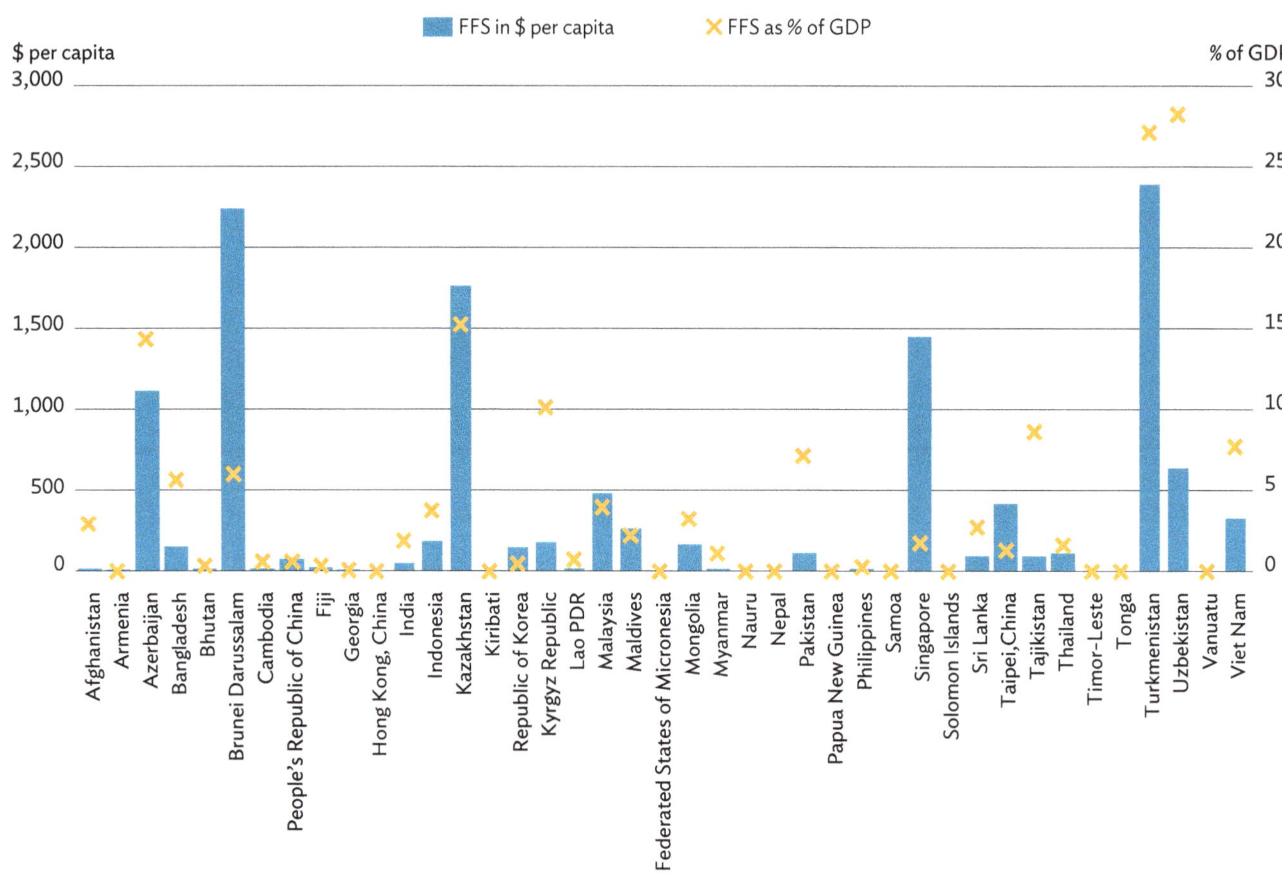

Figure 5.2: Fossil Fuel Subsidies in Developing Asia, 2022

FFS = fossil fuel subsidy, GDP = gross domestic product, Lao PDR = Lao People's Democratic Republic.
Note: Population and GDP data were retrieved from the World Bank Data Repository (2023).
Source: International Institute for Sustainable Development and Organisation for Economic Co-operation and Development. 2024. Fossil Fuel Subsidy Tracker (accessed 29 May 2024).

SOEs, which dominate the supply of energy in developing Asia, are critical to the energy transition.
SOEs account for about 60% of energy investments in developing economies, compared to just 9% in developed economies. They especially dominate investment in fossil fuels and related networks in developing economies, accounting for 59% and 93% of total investment, respectively (IEA 2020). A review of governments in G20 economies found that, in 2022, SOEs invested $348 billion in capital expenditure in coal, natural gas, and oil. The PRC, India, Indonesia, and the ROK accounted for over a third of this investment, totaling more than $146 billion. However, actual SOE investment in fossil fuels is likely much higher due to the lack of a global dataset and poor transparency.

Continued investment in fossil fuel assets is a financial as well as a climate risk. Carbon pricing and "carbon border adjustment"[6] are expected to increasingly impact demand for carbon-intensive products (ADB 2024). Investment in fossil fuels could therefore lead to stranded assets, which is a risk to taxpayers, given these are public assets. Conversely, SOEs can adopt carbon-reduction goals that bring huge investment

---

6   Carbon border adjustment is based on the premise that introducing carbon pricing in a jurisdiction provides an unfair advantage to imports from economies that are not subject to carbon prices. This could lead to higher demand for emissions-intensive imports. To prevent this carbon leakage, an adjustment is imposed on imports to ensure they are subject to a similar carbon price as domestically produced products.

opportunities to benefit both investors and the state (Fang and Jialu 2023). They can develop and implement evidence-based diversification strategies, and increase transparency about their investments, ambitions, and progress. This would be aligned with their institutional mandates, which typically go beyond the profit maximization of private sector counterparts, to include commitments to national energy security and social welfare. SOEs in the PRC, India, and Indonesia have responded to their governments' renewable and net zero ambitions, establishing road maps and investment packages for the deployment and integration of renewable energy sources (Fang and Jialu 2023; Tobing 2022; PLN 2021; Laan, Geddes, Bois von Kursk et al. 2023; Viswanathan et al. 2022).

Despite commitments to reform fossil fuel subsidies, progress has been slow. All developing economies in Asia and the Pacific have pledged to rationalize and phase out inefficient fossil fuel subsidies that encourage wasteful consumption through their participation in Asia-Pacific Economic Cooperation (APEC), the G7, the G20, or the United Nations (APEC 2021; G7 2022; G20 2021; UNDESA n.d.). However, as noted by van Asselt et al. (2023), only 29 of the 198 Parties to the Paris Agreement, including 12 from developing Asia, contain a reference to fossil fuel subsidy reform in their nationally determined contributions. Fossil fuel subsidy reforms have been slow, partly due to loopholes in current international commitments, including a lack of clear timelines.[7] Before the energy crisis initiated by the Russian invasion of Ukraine, several developing Asian economies—notably Bangladesh, India, Indonesia, and Sri Lanka—had made progress on removing fossil fuel subsidies or increasing taxation. Box 5.1 explains relevant achievements in India and Indonesia, although some of this progress has been undermined by the reemergence of subsidies in both countries.

Fossil fuel subsidies tend to disproportionately benefit higher-income consumers. Fossil fuel subsidies for consumers are frequently justified as delivering universal energy access, being used as a form of social support, particularly in economies without alternative social welfare systems, and during periods of high energy prices. During the energy crisis in 2021–2022, many governments tried to shelter consumers from record-high fossil fuel prices, but only some of this support targeted vulnerable consumers (Sgaravatti et al. 2023; IEA 2023). Coady, Flamini, and Sears (2015) find that only $1 of every $10 in gasoline subsidies and $1 of every $30 in liquefied petroleum gas subsidies went to the bottom 40% of households in Asia and the Pacific. Evidence suggests that a more-effective policy to address cost-of-living pressures linked to energy pricing in the short term is through direct cash transfers to vulnerable households and targeted subsidies for critical energy services (Ari et al. 2022). Furthermore, examples in India and Indonesia show that a carefully designed and implemented phaseout of fossil fuel subsidies can result in the redirection of resources to social welfare, health, education, and clean infrastructure.

Governments can tackle subsidy reform by learning from successful international experiences. To act on fossil fuel subsidy reform, governments should develop national targets to align public resources with climate commitments, supported by detailed road maps tailored to local circumstances, evidence-based strategies, and firm deadlines for implementation. International experience in this area indicates that a reform plan should be based on inputs from appropriate government agencies and consultation with relevant stakeholders. A clear communication strategy to publicize the benefits of reform helps build political and public support. The appropriate pace of reform will be based on the type of subsidy, prevailing energy prices, and national circumstances. Finally, pricing mechanisms and SOEs need to be reformed to ensure subsidies do not reemerge (Beaton et al. 2013; Coady et al. 2013).

---

[7]   All current agreements refer to the phaseout of "inefficient" subsidies, but there is no universally endorsed definition of what makes a subsidy "inefficient" (OECD and IEA 2021). In addition, aside from the G7, which has committed to reform "inefficient" subsidies by 2025, all other commitments lack specific deadlines.

Box 5.1: Fossil Fuel Reform in India and Indonesia

India: Shifting Support from Fossil Fuels to Clean Energy

Since about 2010, India has made noteworthy progress on fossil fuel subsidy reform through a calibrated "remove", "target", and "shift" approach (Bridle et al. 2019; Raizada et al. 2024). By carefully balancing the combined effect of three key policy levers—retail prices, tax rates, and subsidies on selected petroleum products—the country was able to reduce its fiscal subsidy in the oil and gas sector by 85%, from an unsustainable peak of $25 billion in 2013 to $3.5 billion in 2023 (Jain 2018; PPAC 2024; Raizada et al. 2024). India gradually phased out the subsidy on petrol and diesel (from 2010 to 2014) and carried out incremental tax increases (from 2010 to 2017), which created fiscal space to increase government support for renewable energy, electric vehicles, and strengthening of electricity infrastructure. The additional tax revenues from increases in excise duty on petrol and diesel from 2014 to 2017, a period of low international crude oil prices, were also redirected to improve access and target subsidies for expanding the use of liquefied petroleum gas (LPG) among the rural poor (Laan, Sharma, and Beaton 2018). Subsidies for LPG have since grown and may now require efforts to improve targeting and to incubate non-fossil-fuel cooking alternatives.

From 2010 to 2017, the Government of India introduced a cess (tax) on coal production and imports. Around 30% of the cess collections were channeled to a national clean energy and environment fund that supported clean energy projects and research (Government of India 2017). The cess significantly contributed to strengthening the budget of India's Ministry of New and Renewable Energy during 2010–2017 and provided the initial funds for the country's Green Energy Corridor scheme and its National Solar Mission, which helped bring down the cost of utility-scale solar energy and fund many off-grid renewable energy solutions. However, with the introduction of the goods and services tax (GST) in India after 2017, the cess on coal production and imports was subsumed within the country's GST compensation cess, the flows of which were redirected to compensate states for revenue losses associated with the new tax regime (Garg et al. 2020).

As a result of India's subsidy reforms and taxation measures, the country's fossil fuel subsidies plummeted from 2014 to 2018, as shown in the figure below. Its renewable energy subsidies also reached a peak in 2017 but are now once again growing, with major support schemes targeting solar parks, state-owned enterprises (SOEs), and distributed renewable energy (Aggarwal et al. 2022).

India's Subsidies for Fossil Fuels and Clean Energy, 2014–2023

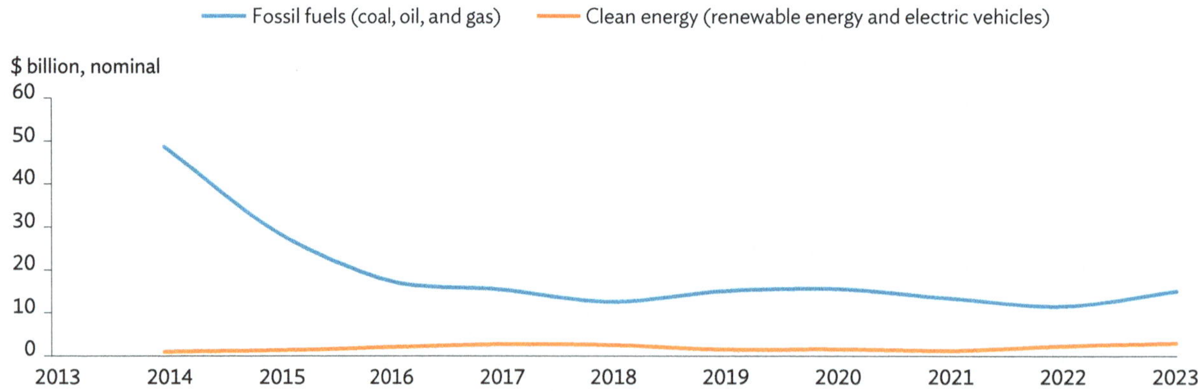

Source: Raizada, S., D. Sharma, T. Laan, and S. Jain. 2024. *Mapping India's Energy Policy 2023: A Decade in Action*. International Institute for Sustainable Development.

Indonesia: Shifting Support from Fossil Fuels to Social Protection

Declining oil production and mounting subsidy bills when fossil fuel prices spiked led the Government of Indonesia to implement major reforms in 2005, 2008, 2013, and 2014. In 2014, Indonesia implemented measures to reform fossil fuel subsidies, taking advantage of falling world oil prices. These measures included the removal of all gasoline subsidies except for distribution costs outside the Java-Madura-Bali area. Price caps on diesel were replaced with a fixed subsidy, allowing diesel prices to fluctuate according to international prices but remaining Rp1,000 per liter below the market price. This policy resulted in a saving for the government of roughly $15.6 billion—over 10% of Indonesia's annual expenditure in 2015.

continued on next page

Box 5.1 *continued*

A comparative analysis conducted by the International Institute for Sustainable Development demonstrates that the savings appear to have been reallocated to regional governments and villages, special programs in health insurance, housing for low-income groups, improved water access for all, and infrastructure investments by SOEs (Pradiptyo et al. 2016).

Despite Indonesia's efforts to reform its policies on fossil fuels, subsidy levels have surged again. From 2016 to 2020, the number of support measures for the Indonesian energy sector steadily increased, with a sharp rise in 2020, largely attributed to economic recovery packages that were mostly directed toward bailing out SOEs (EPT 2024). Most of Indonesia's support for fossil fuels comes from the government compensating the SOE Pertamina for the low retail prices it sets for gasoline, diesel, and some sales of LPG and kerosene (OECD 2023; Suharsono et al. 2022). This compensation rose sharply in response to the global energy crisis in 2021–2022, which induced a sharp increase in global prices of fossil fuels. As shown in the figure below, Indonesia spent around $50 billion in 2022 to support fossil fuels, an increase of nearly $35 billion on the amount spent in 2020 (IISD and OECD 2024).

Indonesia's Fossil Fuel Subsidies, 2010–2022

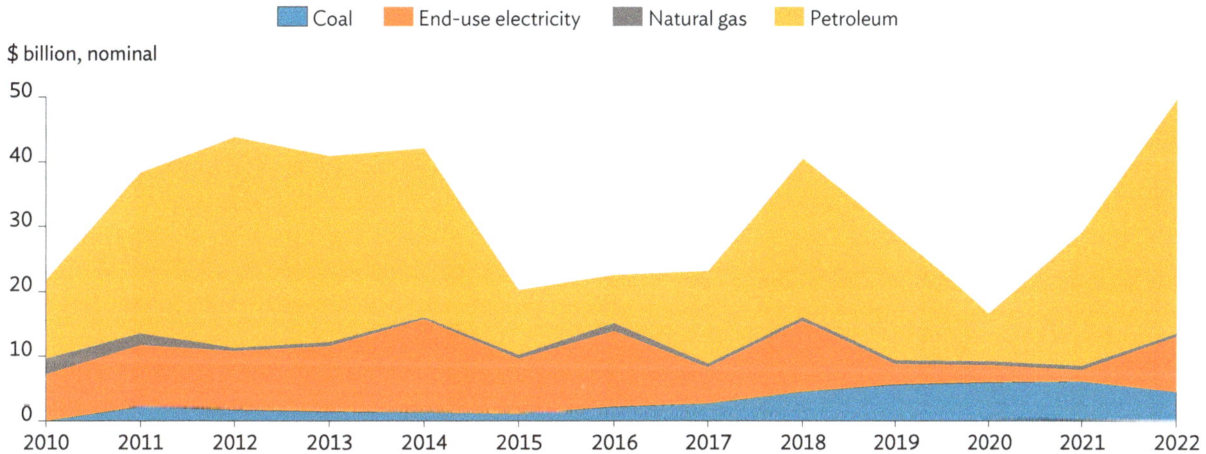

Source: International Institute for Sustainable Development and Organisation for Economic Co-operation and Development. 2024. Fossil Fuel Subsidy Tracker (accessed 29 May 2024).

References:
Aggarwal, P., S. Goel, T. Laan, T. Mehta, A. Pant, S. Raizada, B. Viswanathan, A. Viswamohanan, C. Beaton, and K. Ganesan. 2022. Mapping India's Energy Policy 2022. International Institute for Sustainable Development (IISD) and the Council on Energy, Environment and Water.
Bridle, R., S. Sharma, M. Mostafa, and A. Geddes. 2019. Fossil Fuel to Clean Energy Subsidy Swaps: How to Pay for an Energy Revolution. IISD.
EPT (Energy Policy Tracker). 2024. Indonesia (accessed 19 April 2024).
Garg, V., B. Viswanathan, D. Narayanaswamy, C. Beaton, K., Ganesan, S. Sharma, and R. Bridle. 2020. Mapping India's Energy Subsidies 2020: Fossil Fuels, Renewables, and Electric Vehicles. IISD.
Government of India (Ministry of Finance, Department of Expenditure). 2017. National Clean Energy & Environment Fund Brief (updated for actuals of fiscal year 2017 and fiscal year 2018 from Union Budget documents).
IISD (International Institute for Sustainable Development) and OECD (Organisation for Economic Co-operation and Development). 2024. Fossil Fuel Subsidy Tracker (accessed 29 May 2024).
Jain, A. K. 2018. A Fine Balance: Lessons from India's Experience with Petroleum Subsidy Reforms. Energy Policy. 119. pp. 242–249.
Laan, T., S. Sharma, and C. Beaton. 2018. Support for Clean Cooking in India: Tracking the Latest Developments in LPG Subsidies. IISD.
OECD. 2023. OECD Companion to the Inventory of Support Measures for Fossil Fuels: Country Notes—Indonesia.
PPAC (Petroleum Planning and Analysis Cell, Ministry of Petroleum & Natural Gas, Government of India). 2024. Snapshot of India's Oil and Gas Data: Monthly Ready Reckoner January 2024.
Pradiptyo, R., A. Susamto, A. Wirotomo, A. Adisasmita, and C. Beaton. 2016. Financing Development with Fossil Fuel Subsidies: The Reallocation of Indonesia's Gasoline and Diesel Subsidies in 2015.
Raizada, S., D. Sharma, T. Laan, and S. Jain. 2024. Mapping India's Energy Policy 2023: A Decade in Action. IISD.
Suharsono, A., M. Hendriwardani, T. B. Sumarno, J. Kuehl, M. Maulidia, and L. Sanchez. 2022. Indonesia's Energy Support Measures: An Inventory of Incentives Impacting the Energy Transition. p. 51. IISD.

Source: Do, N., T. Laan, S. Raizada, L. Cameron, D. Sharma, O. B. von Kursk, and C. Beaton. Fossil Fuel Subsidy Reform. Background Paper for the *Asia–Pacific Climate Report 2024: Catalyzing Finance and Policy Solutions*. Asian Development Bank.

Reforms need to be complemented by policies to protect vulnerable consumers. Policy reforms around fuel subsidies may not be politically viable unless they also address a wide array of social goals. Removing subsidies can sometimes cause hardship for vulnerable populations and businesses and below-market fossil fuel prices are often seen as part of the social contract in many energy-producing economies (McCulloch 2023). Governments therefore need to identify vulnerable groups and businesses, anticipate impacts, and develop targeted assistance measures and compensation (such as cash transfers funded by subsidy savings) to address these impacts. In addition, some temporary financial support for the middle classes may also be needed to reduce the likelihood of political backlash. A commitment to reinvest subsidy savings in education, health, or cash transfers has been seen to double public support for subsidy reform (World Bank 2023a).

5.2.2. Agriculture Subsidies to Be Replaced by Climate-Smart Investments

Subsidies for agriculture and land use drive emissions and environmental degradation. After the energy sector, agriculture and land-use change combine as the second-largest source of GHG emissions, accounting for a quarter of all emissions worldwide. Globally, agriculture subsidies reached a historical high of $851 billion per year for 2020–2022, a 2.5-fold increase from the levels of 2000–2002 (OECD 2023). The PRC and India accounted for 36% and 15%, respectively, of global agriculture subsidies in 2020–2022. About 74% of global agriculture subsidies go to producers in the form of market price support (such as import tariffs and other border measures) or payments for use of inputs (such as fertilizers and improved seeds). In parts of Asia, agriculture input subsidies have promoted such excessive overuse of fertilizers that yields have fallen and water pollution through increased nitrogen runoff has increased (World Bank 2023b). Similarly, subsidies that lower the price of electricity, diesel, and water for irrigation have promoted overuse and led to methane emissions, especially from rice cultivation (Rahut et al. 2023; Giardino, Wilson, and Hart 2023). Rice receives the highest share of subsidies among all commodities and is one of the top three sources of agriculture's GHG emissions (OECD 2023). In some economies of developing Asia, state concessions in forest areas have effectively subsidized deforestation and increased both emissions and environmental degradation (ADB 2023a).

Agriculture subsidies are poorly targeted and exacerbate existing inequalities. Most agriculture subsidies accrue to wealthier farmers, even when programs are designed to target the poor (Ricker-Gilbert and Jayne 2016; World Bank 2023b). Consumer subsidy programs usually require significant out-of-pocket expenditure by recipients, which discourages participation from lower-income households. Meanwhile, producer subsidies tend to scale with levels of production, which means a disproportionate share goes to larger producers.

Reforming and repurposing agriculture subsidies can help farmers adapt to climate change. A majority of agriculture subsidies are concentrated in market-distorting measures such as market price support and use of inputs, while the share of subsidies that support research and development, extension services, innovation, and infrastructure is small and declining (OECD 2023). Investment in these areas has the potential not only to reduce emissions and environmental degradation but also to enhance the ability of farmers to adapt to the changing climate. Similar to the scenario with fossil fuels, agriculture subsidy reform should be based on a transparent process and multistakeholder consultation, with a clear communication strategy. It should also entail measures for those who face losses, such as cash transfers that target vulnerable people (OECD 2023; FAO, UNDP, and UNEP 2021).

5.3 Carbon Taxation as a Core Carbon Pricing Policy

In Asia and the Pacific in 2024, only Japan and Singapore had implemented a national carbon tax.
Japan began imposing a carbon tax in 2012, based on CO_2 emissions from the combustion of fossil fuels
across all sectors. Japan's carbon price has been steady at ¥289 per metric ton of CO_2 equivalent (tCO_2e)
since 2016. However, the Government of Japan provides exemptions and refund measures for specific fossil
fuels used in the agriculture, energy, industry, and transport sectors. Singapore started imposing a carbon
tax in 2019, targeting facilities in the energy, manufacturing, waste, and water sectors that directly emit at
least 25,000 tCO_2e per year. Singapore has followed a phased approach to increasing carbon prices. Initially
set at S$5 per tCO_2e in 2019, the price increased to S$25 in 2024, and is set to increase further to S$45 by
2027 and to as much as S$80 by 2030. In Japan and Singapore, revenues collected from carbon taxes will
fund government decarbonization efforts, including investments in low-carbon technologies and practices
and implementation of energy efficiency measures. In Singapore, revenues will also be used to cushion the
impact of the carbon tax on businesses and households. From 2024, companies may use eligible international
carbon credits to offset up to 5% of their taxable emissions. While carbon taxation is under consideration by
more governments in the Asia and Pacific region,[8] questions remain regarding policy design, macroeconomic
implications, and impacts of a carbon tax on an economy's international competitiveness.

5.3.1 Carbon Tax Policy Design

Carbon taxation presents a compelling strategy for cost-effectively reducing GHG emissions. By letting
polluters decide how to reduce their emissions, a carbon tax ensures that the reduction methods chosen
are those that minimize costs, making it an economically efficient solution. Another primary argument for
carbon taxation is its ability to raise substantial government revenue. In some cases, it can even be a more
efficient revenue source than other types of taxation, making it a worthwhile policy consideration even if GHG
emissions were not a concern. The revenue generated can be used to support public goods and services or to
offset other taxes.

When designing a carbon tax, several elements must be considered. These elements include the tax rate, its
coverage, use of the revenue, and the point of enforcement. The economic principle for setting the rate of a carbon
tax is to align it with the harm caused by GHG emissions. Recent estimates suggest that the cost of the marginal
damage from each extra metric ton of GHG emissions is around $200 (Rennert et al. 2022), although there is
considerable uncertainty and controversy over this figure. Prior to that, the best estimate was about $40 per metric
ton. This significant shift highlights the evolving understanding of how GHGs impact the economy.

Even low rates of carbon taxation can deliver significant environmental benefits. Despite the ideal of
setting the carbon tax rate at $200 per metric ton, such a rate may be politically challenging, especially in
low-and middle-income economies. Therefore, alternative approaches could include setting a tax rate to meet
specific emissions targets, such as those in the Paris Agreement, or coordinating tax rates with neighboring
nations and major trading partners. The most realistic approach, however, is setting a tax rate based on
what is politically feasible, because even relatively low tax rates can lead to substantial emissions reductions.
For example, a modeling exercise shows that a carbon tax of $12 to $15 per metric ton could achieve emissions
cuts similar to those expected from the Inflation Reduction Act in the United States (Bistline, Mehrotra,
and Wolfram 2023).

---

[8]   Taipei,China is set to impose a carbon fee in 2024, which targets electricity and manufacturing facilities that emit more than
      25,000 tCO_2e of GHG. The government has not set the fee rate but has announced that preferential pricing will be provided to
      entities with approved emissions reduction plan.

In principle, all GHG emissions should be taxed based on their relative warming potential. Process emissions, such as those from cement production, and emissions of non-CO_2 gases such as methane and nitrous oxide, are hard to monitor and measure. Land-use changes, such as deforestation, also present significant challenges. Given these difficulties, most economies have initially focused on energy-related CO_2 emissions. This focus poses dilemmas for economies where other sources are significant contributors of emissions. For example, in Indonesia, land-use change is a major emissions source that needs addressing despite the monitoring challenges.

Economic reasoning suggests that revenue from carbon taxation should be redirected to the highest-value public uses. Such uses might include offsetting distortionary taxes, compensating households through direct rebates or tax cuts, supporting affected industries, addressing trade and competitiveness issues, funding infrastructure development, or financing new environmental initiatives. The optimal use depends on specific context of an economy. In any case, the carbon tax revenue should not be treated differently from other public revenues just because it comes from protecting the environment.

The easiest method for enforcing a carbon tax is to tax fossil fuels upstream. This method involves taxing coal, oil, and natural gas at natural "choke points" in the supply chain, such as import points or refineries. This approach simplifies enforcement by focusing on a small number of entities, minimizing opportunities for tax evasion and avoidance. It has considerable advantages for lower- and middle-income economies in particular, as these economies usually have large informal sectors and limited enforcement capacity. However, this approach only works for energy-related CO_2. Other emissions, such as process emissions and non-CO_2 gases, must be enforced downstream, where the actual emissions occur, which complicates enforcement.

5.3.2   Challenges and Design Responses of Carbon Tax Policy

Implementing a carbon tax policy presents several significant challenges. The most prominent of these challenges are concerns around distributional impacts, uncertainty on emissions reductions, and international trade and competitiveness (covered later in this chapter). These issues need careful consideration and strategic responses to ensure the policy's effectiveness and fairness.

A carbon tax may have regressive effects, potentially harming low-income households. This scenario is largely due to the increased costs of energy goods, which are often necessities. However, the regressive nature of a carbon tax may be less severe than initially perceived. Many of the effects of a carbon tax occur through the prices of other goods, some of which are much less regressive. For instance, higher prices for air travel primarily impact wealthier consumers. Moreover, even the energy price effects may be less regressive in developing economies than in the developed economies because what is considered a necessity, such as cars in the United States, might be viewed as a luxury in lower- and middle-income jurisdictions. Meanwhile, the income effects of a carbon tax might be progressive. Evidence from high-income economies indicates that a carbon tax tends to reduce capital income more than labor income, and capital income is mainly accrued by higher-income households. This progressive income effect can offset some of the regressive impacts on consumption.

The potentially regressive impacts of a carbon tax can be balanced by appropriate use of revenue. The first option under such a strategy is through per capita rebates that distribute the revenue generated from the carbon tax equally among all citizens. Another option is income-targeted transfers, i.e., direct financial assistance to low-income households to offset increases in energy costs. Tax revenues can also be used to fund anti-poverty initiatives that support low-income communities. Last, reducing other regressive taxes can also help alleviate the financial burden on vulnerable households. Concerns about specific industries or firms can be addressed through output-based rebating or marginal exemptions. These measures might reduce policy efficiency slightly, but they can effectively mitigate undesirable distributional effects at a relatively low cost.

Imposing a carbon tax may not provide certainty on emissions reductions. Carbon taxes provide price certainty, but they leave emissions levels variable. If emissions certainty is a major concern, an adjustment mechanism can be introduced whereby tax rates are adjusted according to a target emissions path. If emissions are above target, the carbon tax rate is increased and vice versa. Although this approach sacrifices some price certainty, it helps stabilize emissions by creating a hybrid policy between a fixed carbon tax and an emissions trading system (ETS).

5.3.3 Macroeconomic Implications of a Carbon Tax

It has been mooted that a carbon tax could hinder economic growth. When considering adoption of a carbon tax, a key question is the potential impact on the broader economy, particularly on economic growth, inflation, and employment. Understanding these macroeconomic implications is crucial for designing an effective and sustainable carbon tax policy.

Economic analysis suggests the economic impacts of a carbon tax are moderate. Model-based simulations indicate that the impact of a carbon tax on economic growth is likely to be small and could even be slightly positive (Goulder and Hafstead 2013; Lu, Tong, and Liu 2010; McKibbin et al. 2015). The effect tends to be more positive if the carbon tax is applied upstream, meaning at the point of fossil fuel extraction or import, and if the revenue generated is used to reduce other inefficient taxes. The primary impact of a carbon tax is not to reduce growth but to shift the economy from high-carbon to low-carbon activities, thereby affecting the carbon intensity of the economy rather than overall gross domestic product. Empirical research on existing carbon pricing strategies shows that established mechanisms have not significantly impacted economic growth rates once implemented (Köppl and Schratzenstaller 2023). Studies consistently find no large positive or negative effects on economic growth, with confidence intervals centered around zero, indicating minimal impact (Bernard, Kichian, and Islam 2018; Metcalf and Stock 2020). However, it should be noted that results may change if evaluations are done over longer time frames, focus on high rates of carbon taxation, and/or take into account international interactions. For instance, modeling of global economic impacts up to 2050 shows that boosting green investment using a combination of climate mitigation policies, including carbon pricing, can increase global output for the first 15 years, but output reverts below baseline levels once the impacts associated with stimulating investments wear off (Jaumotte, Liu, and McKibbin 2021).

Another concern is the potential for a carbon tax to drive inflation. However, evidence suggests that there are no significant inflationary effects following the implementation of a carbon tax. This finding is somewhat surprising, as economic models typically suggest that a carbon tax would pass through to consumer prices, potentially raising overall price levels. Research by Konradt and Weder di Mauro (2023) indicates that potential price increases are either offset by other factors or do not materialize as expected, leading to no significant change in inflation. The study suggests that proper redistribution of revenue and appropriately calibrated central bank interventions can neutralize any inflationary pressure from carbon pricing policies.

A key issue is that a carbon tax can impact employment, particularly in energy-intensive industries. While there may be job losses in carbon-intensive industries, these are often offset by employment gains in low-emissions sectors, resulting in a net shift rather than a net loss of jobs (Yamazaki 2017; Shapiro and Metcalf 2023). Nevertheless, this job shift can pose challenges for workers in affected industries. To mitigate these effects, it is advisable to preannounce or phase in the carbon tax, allowing more time for the labor market to adjust. Additionally, implementing policies that provide compensation or retraining for workers in impacted sectors can help ease the transition.

5.3.4   Wider Adoption of Carbon Taxation

It is vital that politicians and the general public be united in their climate ambitions. Perhaps the most formidable challenge facing the widespread adoption of a carbon tax is securing sufficient political will and gaining public acceptance. Unlike technical feasibility, political and societal readiness can be unpredictable. These challenges can be navigated through strategically planned measures such as rebates to vulnerable households, transition assistance for affected industries, clear and transparent preannouncements on policy, and gradually phasing in the tax regime. However, despite these efforts, significant resistance from various stakeholders can remain, which is when implementing an ETS becomes a viable alternative for many economies, as discussed in Box 5.2.

Box 5.2: Carbon Taxation Versus Emissions Trading System

A carbon tax and an emissions trading system (ETS) are two main carbon pricing policy instruments. In an ideal world, they are equivalent in efficiently achieving environmental objectives. However, in practice, there are some notable differences between the two policies.

The first major advantage of carbon taxation over an ETS is the price certainty. A carbon tax provides a clear and predictable price path for greenhouse gas (GHG) emissions. This predictability is crucial for businesses as they plan long-term investments. Knowing the future cost of GHG emissions allows companies to make informed decisions without the risk of facing volatile prices that could drastically change in a short period. This stability is especially valuable for capital-intensive industries where investment horizons span several years.

Another key benefit of a carbon tax is its relative simplicity and transparency. Unlike an ETS—which requires complex trading mechanisms, continuous cap-setting, allocation mechanisms, and monitoring of trades—a carbon tax is straightforward to implement. It clearly outlines the costs associated with GHG emissions, making the financial flows easy to track and understand. This transparency simplifies the process for both the government and businesses, reducing administrative burdens and enhancing public trust in the system.

A carbon tax can also generate substantial revenue for the government and this revenue can be used to fund various public services or offset other taxes. While ETSs can also generate revenue if permits are auctioned, this is not inherently built into the system: in reality, most emissions permits are given to emitters for free.

Furthermore, the implementation of a carbon tax via the relevant tax authority offers a distinct advantage in lower- and middle-income economies. By leveraging the existing infrastructure and expertise of tax authorities, a carbon tax can be more effectively enforced, ensuring better compliance and outcomes. By contrast, environmental ministries in these economies may lack the necessary capacity and resources to build and regulate an ETS.

However, ETSs do have several notable advantages over carbon taxation, particularly in terms of emissions certainty. An ETS provides a clear limit on emissions by setting a cap, allowing the market to determine the price of emissions permits. This approach ensures that total emissions remain within a predefined limit, which can be set to meet specific environmental targets.

Distributional flexibility is another significant advantage of an ETS. The allocation of permits can be adjusted to support particular groups or industries, allowing targeted economic support while still adhering to overall emissions limits. While similar effects can be achieved with tax rebates in a carbon taxation system, the inherent design of an ETS makes such adjustments more straightforward. Additionally, the allocation of permits can be done discreetly, without drawing public attention to the redistribution of resources. This can be advantageous in situations where explicit financial transfers might be politically sensitive or controversial.

However, broadly speaking, these two carbon pricing policies are more similar than they are different, with both approaches sharing the common goal of reducing GHG emissions and mitigating climate change at the lowest economic cost. Design and implementation features probably matter more than the carbon taxation versus ETS debate.

Source: Authors.

5.4 Build and Connect Emissions Trading Systems

5.4.1. Emissions Trading System Practices in Asia and the Pacific

Of the 13 national ETSs established worldwide by 2024, six are in the Asia and Pacific region.[9] National ETSs of various types operate in Australia, the PRC, Indonesia, Kazakhstan, the ROK, and New Zealand. New Zealand's ETS dates back to 2008 and is the second oldest in the world.[10] It was followed by Kazakhstan and the ROK in 2013 and 2015, respectively. By 2016, the PRC had initiated pilot ETSs in eight cities[11] and established a nationwide ETS in 2021. Australia does not have a formal ETS, but has operated a mandatory baseline-and-credit system since 2023.[12] Indonesia's ETS commenced in 2023.

The region's six national ETSs differ in terms of their ambitions and implementation. All of them, however, are central to meeting the GHG mitigation commitments of their respective countries: to achieve net zero by 2050 for Australia, the ROK, and New Zealand; and by 2060 for the PRC, Indonesia, and Kazakhstan (Table 5.1). In addition, all of these economies are also aiming for significant GHG reductions by 2030. This is noteworthy especially for the PRC, Indonesia, Japan, the ROK, and Australia, which are the world's first, sixth, seventh, 13th, and 15th largest emitters of GHGs, respectively (European Commission 2024). In terms of scope and coverage, the ROK's ETS is the most comprehensive, as it includes all of the GHGs nominated in the Kyoto Protocol. New Zealand's ETS is also comprehensive, though it excludes biogenic methane, which is a major emissions category for that jurisdiction. Most of the other national ETSs focus primarily on CO_2 emissions. The share of total emissions covered by ETSs ranges from a low of 26%–29% (Australia and Indonesia) to a high of 88.5% (ROK). The ROK and New Zealand have the most comprehensive coverage in terms of sectors, while the other systems mostly focus on the energy sector, a major source of CO_2 emissions.

There is very little auctioning of emissions allowances in any ETS in Asia and the Pacific. Allowances within the region's ETSs are mostly given for free, especially for emissions-intensive and trade-exposed industries. In New Zealand's ETS, 54% of allowances were offered for auction in 2022, but none was sold, since the clearing price was lower than the reserve price (ICAP 2022). With regard to allowance markets, only two ETSs (the ROK and New Zealand) had a long enough time series for prices.[13] Allowance prices have been low in the ROK and moderate in New Zealand, especially as compared to the ETS of the European Union (EU) as shown in Figure 5.3. Additionally, in line with the experience of the EU's ETS, allowance prices in the ROK and New Zealand have shown volatility. In the latter, prices fell to practically zero ($2) in 2013—soon after the announcement delinking the New Zealand ETS from Kyoto markets—but prices have been mostly rising since then.

---

[9]   The 13 national ETSs worldwide are found in Australia, Austria, Canada, the PRC, Germany, Indonesia, Kazakhstan, the ROK, Mexico, Montenegro, New Zealand, Switzerland, and the United Kingdom.

[10]   The first ETS for GHGs was the European Union's ETS in 2005. Box 5.3 provides a brief overview of the EU's ETS.

[11]   The eight cities are Beijing, Chongqing, Fujian, Guangdong, Hubei, Shanghai, Shenzhen, and Tianjin.

[12]   Unlike an ETS, in a baseline-and-credit system there are no emission caps. Firms below a given baseline can generate and sell credits to other firms that exceed the baseline.

[13]   The PRC's ETS market is nascent, but data for 2 years are shown in Table 5.1. Kazakhstan's ETS had stops and starts and allowance prices are not significantly different from zero (Table 5.1, row 10).

Table 5.1: Main Features of National Emissions Trading Systems in Asia and the Pacific

		Australia	PRC	Indonesia	Kazakhstan	ROK	New Zealand
1.	GHG reduction targets for the country	By 2030: 43% below 2005 levels By 2050: Net zero emissions	By 2030: 65% below 2005 levels By 2060: Carbon neutrality	By 2030: 31.9% below BAU By 2060: Climate neutrality	By 2050: 40% reduction in energy sector emissions from 2012 levels. By 2060: Carbon neutrality	By 2030: At least a 35% reduction below 2018 emissions By 2050: Carbon neutrality	By 2030: 50% reduction of net emissions below gross 2005 levels By 2050: Reduce net emissions of all GHGs (except biogenic methane) to zero
2.	Year ETS introduced	2016	2021	2023	2013	2015	2008
3.	GHGs covered	CO_2, CH_4, N_2O, SF_6, HFCs, PFCs	CO_2	CO_2, CH_4, N_2O	CO_2	CO_2, CH_4, N_2O, SF_6, HFCs, PFCs	CO_2, CH_4, N_2O, SF_6, HFCs, PFCs (but no biogenic CH_4)
4.	Covered emissions (% total emissions and year)	26% (2020)	40% (2021)	29% (2023)	47% (2021)	88.5% (2021)	48% (2021)
5.	Emissions subject to ETS	137 $MtCO_2e$	5,200 $MtCO_2e$ (approx.)		160 $MtCO_2e$	599 $MtCO_2e$	35.9 $MtCO_2e$
6.	Sectoral coverage	Waste, Domestic Aviation, Transport, Industry	Power	Power	Industry, Energy	Waste, Domestic Aviation, Transport, Buildings, Industry, Energy	Forestry, Maritime, Waste, Domestic Aviation, Transport, Buildings, Industry and Energy
7.	No. of obligated entities/firms (year)	215 (2023)	2257 (2021–2022)	63 entities, 146 installations (2024)	128 companies (201 installations)	804 (2023)	4114 (2023)
8.	Allowance allocation	Set up on the basis of production levels multiplied by an emission intensity value	Output based benchmarking for 4 benchmarks: conventional coal plants below 300 MW; above 300 MW; unconventional coal and natural gas. Free allocation currently.	Auctioning, and benchmarking	Free allocation (benchmarking and grandparenting approach)	Free allocation (benchmarking and grandparenting approach) and auctioning	Free allowance and auction Note: Proportion of allowances offered at auction in 2023 was 54% but none were sold (winning bid less than reserve price).
9.	Compliance period	Annual	Annual	Annual	Annual	Annual	Annual
10.	Average prices (2023)		Secondary market: CNY68.35 ($9.65)	Auction: Rp10,000 ($0.64) Secondary market: Rp69,600 ($4.45)	Secondary Market: T473 ($1.04)	Auction: W10,672 ($ 8.17) Secondary market: W9,999 ($7.66)	Auction: NZ$0 ($0) Secondary market: NZ$62.79 ($38.30)

continued on next page

Table 5.1 *continued*

	Australia	PRC	Indonesia	Kazakhstan	ROK	New Zealand
11. Monitoring, reporting, and verification	Monitoring: Facilities need to apply for a site-specific emissions intensity value for the first compliance year. Reporting: Annual self-reporting according to the guidelines defined in the "National Greenhouse and Energy Reporting Act." Verification: All facilities with emissions greater than 1 MtCO$_2$e per year are required to undergo independent emissions audit each year.	Monitoring: Covered entities are required to set up monitor plans and monitor their emission based on these plans. Reporting: Covered entities must submit the previous year's emissions reports by the end of April each year. Verification: Provincial-level ecological and environmental authorities are responsible for organizing the verification of GHG reports. Verification of emissions from the energy sector must be complete by the end of June.	Monitoring: An MRV system is currently in operation in the industry sector and the power generation sub-sector. Pilot MRV programs are being conducted also in the cement and fertilizer sectors. Reporting: The reports must be submitted by the end of January for the preceding year. Installations must report emissions of considered GHGs expressed in units of CO$_2$e. Verification: It should be completed by the end of March, following the January reporting deadline. Verifiers are required to adhere to the guidelines for GHG emission verification in the power subsector.	Reporting: Required annually for installations above the 20,000 MtCO$_2$e/year threshold. Verification: Emissions data reports and their underlying data require third-party verification by an accredited auditor.	Reporting: Annual reporting of emissions from the previous year must be submitted by the end of March. Verification: Emissions must be verified by a third-party verifier. Framework: Emission reports are reviewed and certified by the Certification Committee of the Ministry of Environment by the end of May.	Reporting: Most sectors are required to report annually; the deadline is the end of March to submit an annual emission return. Verification: follows a system of self-reporting supplemented by a program of official government audits. Third-party verification is not required for emissions reports. However, participants must seek third-party verification if they apply for the use of a unique emissions factor, as opposed to using the default factors supplied by the government.

BAU = business as usual; PRC = People's Republic of China; CH$_4$ = methane; CO$_2$ = carbon dioxide; CO$_2$e = carbon dioxide equivalent, ETS = emissions trading system; GHG = greenhouse gas; HFC = hydrofluorocarbon; MRV = monitoring, reporting, and verification; MtCO$_2$e = metric tons of carbon dioxide equivalent; MW = megawatt; N$_2$O = nitrous oxide; PFC = perfluorocarbon; ROK = Republic of Korea; SF$_6$ = sulfur hexafluoride.
Source: Gupta, S., S. Basu-Das, and S. Gupta. 2024. Carbon Pricing Through Emissions Trading In Asia-Pacific: A Comparative Analysis and the Way Forward. Background paper for the *Asia–Pacific Climate Report 2024: Catalyzing Finance and Policy Solutions*. Asian Development Bank.

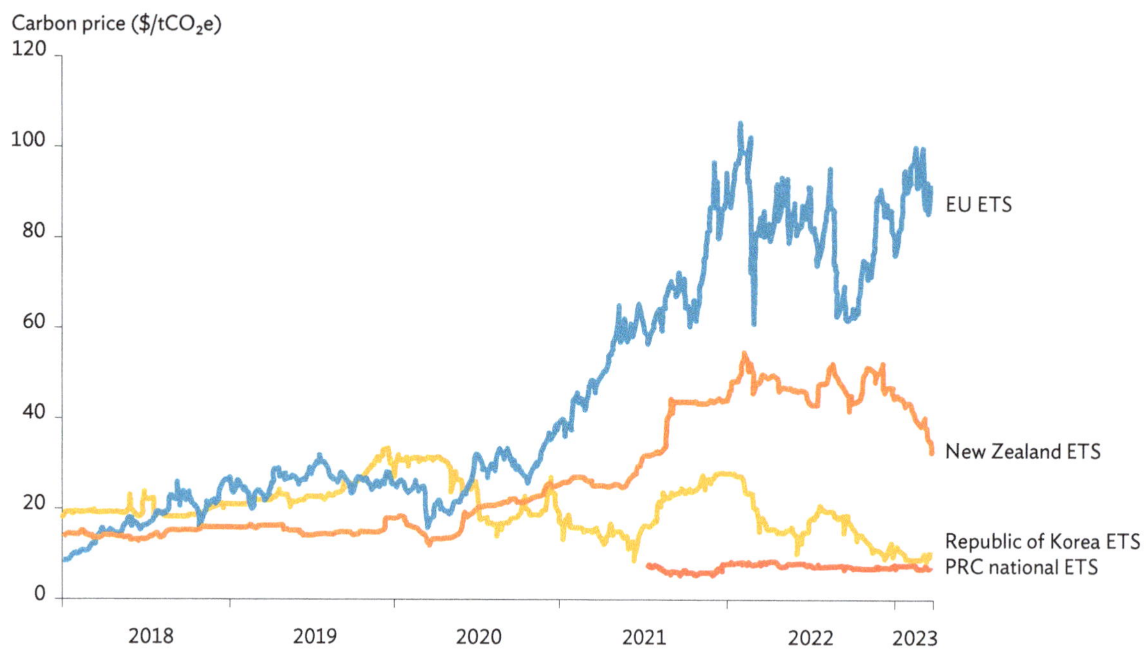

PRC = People's Republic of China, ETS = emissions trading system, EU = European Union, tCO_2e = metric ton of carbon dioxide equivalent.
Source: World Bank. 2023. State and Trends of Carbon Pricing 2023.

5.4.2   Lessons for Economies in Asia and the Pacific

Developing an ETS should be an exhaustive process, with pilot projects and policy flexibility. This approach can be observed in the case of the PRC, which undertook ETS experimentation during the early 2010s, yielding pilot systems that were diverse and instructional. These early ETSs spanned the political and business hubs of Beijing and Shanghai, the sprawling industrial municipalities of Tianjin and Chongqing, the manufacturing locus of Guangdong province, the iron and steel center of Hubei province, and the special economic zone of Shenzhen. Based on the pilot schemes, the PRC's State Council released guidelines for accelerating the establishment of a national carbon market in 2017 and the nationwide ETS was introduced in 2021. With guidelines again updated in 2024, the PRC's national ETS is becoming a more impactful policy instrument (Gupta, Basu-Das, and Gupta 2024). Economies in Asia and the Pacific can learn from the PRC's progressive implementation and ever-evolving ETS, and some are taking similar approaches to increasing scope and ambition over time. For instance, Thailand's Greenhouse Gas Management Organisation has operated a voluntary ETS since 2013, focusing on the development and testing of monitoring, reporting, and verification (MRV) systems, setting caps and allocation procedures, and establishing trading infrastructure for 12 sectors with high GHG emissions. Meanwhile, Indonesia and Viet Nam both have three-phase plans that expand in coverage, compliance obligations, and potential impact through the 2020s. Prospective carbon markets elsewhere in South Asia and Southeast Asia will likewise go through progressive development stages if they are enacted.

The benefits and challenges of efficiency-based systems should be closely observed. The PRC, for example, has deployed efficiency-based targets in its ETS because such targets can be adapted to higher levels of economic production and aligned with broader PRC climate targets that lack an overall cap. Indonesia initiated an efficiency-based ETS in early 2023, focusing on power sector facilities with over 100 megawatts of production capacity. India's Perform, Achieve and Trade scheme may evolve into a compliance-based ETS and is also based on efficiency targets. It is likely that other developing economies in Asia and the Pacific will pursue this route, given their growth ambitions. However, while the impetus for efficiency-based targets is clear,

Box 5.3: Empirics of the European Union's Emissions Trading System

The emissions trading system (ETS) of the European Union (EU), a key instrument for decarbonizing the global economy, remains important to the learning experience across Asia and the Pacific.

The EU's ETS remains the world's largest cap-and-trade system, regulating greenhouse gas (GHG) emissions at over 11,000 installations—despite the ETS of the People's Republic of China being bigger in volume. The EU's ETS is also the oldest cap-and-trade system and the only multinational ETS comprising 30 individual economies (27 EU member states plus Iceland, Liechtenstein, and Norway). It is comprehensive in scope and covers about 45% of the EU's GHG emissions and is in its fourth trading phase (2021–2030). The EU's ETS was revised in 2023 in the context of the *European Green Deal* to align it with the target of reducing net emissions by 2030 to 62% below 2005 levels and net zero by 2050.

Since its launch in 2005, the EU's ETS has helped drive down emissions from electricity and heat generation and industrial production by 37.3% (European Commission 2023). It has also generated €184 billion ($206 billion) in auction revenues for distribution to member states of which €43.6 billion was in 2023 alone (ICAP 2024). Member states have mostly used this revenue to fund investments in energy transformation and decarbonization as well as social measures in support of the green transition (European Commission 2023).

A growing body of literature points to significant positive impacts of the EU's ETS, based on economy-specific microdata studies (Colmer et al. 2022; Klemetsen, Rosendahl, and Jakobsen 2020; Petrick and Wagner 2014) and on sector-level data (Bayer and Aklin 2020). Further, a recent cross-study of installation-level data in France, the Netherlands, Norway, and the United Kingdom found that introduction of the EU's ETS was associated with a statistically significant reduction of carbon dioxide emissions in the order of –10% in the first two trading phases from 2005 to 2012. In addition, firm-level data on all ETS-regulated economies show the EU's ETS has had no significant impact on profits or employment levels and has led to an increase in revenues and fixed assets for regulated firms (Dechezleprêtre, Nachtigall, and Venmans 2023).

References:

Bayer, P. and M. Aklin. 2020. The European Union Emissions Trading System Reduced CO_2 Emissions Despite Low Prices. *Proceedings of the National Academy of Sciences.* 117 (16). pp. 8804–8812.

Dechezleprêtre, A., A. Nachtigall, and F. Venmans. 2023. The Joint Impact of the European Union Emission Trading System on Carbon Emissions and Economic Performance. *Journal of Environmental Economics and Management.* 118 (102758). pp. 1–41.

Colmer, J., R. Martin, M. Muuls, and U. Wagner. 2022. Does Pricing Carbon Mitigate Climate Change? Firm-Level Evidence from the European Union Emissions Trading Scheme. *CEPR Discussion Paper.* No. 16982. Centre for Economic Policy Research.

European Commission. EDGAR: Emissions Database for Global Atmospheric Research. GHG Emissions of all World Countries: 2023 Report.

European Commission. 2023. *Report from the Commission to the European Parliament and the Council on the Functioning of the European Carbon Market in 2022.*

ICAP (International Carbon Action Partnership). 2024. EU Emissions Trading System (accessed 10 July 2024).

Klemetsen, M., K. E. Rosendahl, and A. L. Jakobsen. 2020. The impact of the EU ETS on Norwegian Plants' Environmental and Economic Performance. *Climate Change Economics.* 11 (1).

Petrick, S. and U. J. Wagner. 2014. The Impact of Carbon Trading on Industry: Evidence from German Manufacturing Firms. *Kiel Working Paper.* No 1912. Kiel Institute for the World Economy.

Source: Gupta, S., S. Basu-Das, and S. Gupta. 2024. Carbon Pricing Through Emissions Trading In Asia-Pacific: A Comparative Analysis and the Way Forward. Background paper for the *Asia–Pacific Climate Report 2024: Catalyzing Finance and Policy Solutions.* Asian Development Bank.

such targets may limit the ability of an ETS to significantly curb absolute emissions. ETS designs should therefore seek to reduce these risks through broad coverage and steadily tightening benchmarks. Economies should incorporate clear timelines and strategies for ensuring the emissions impacts of their ETSs and for transitioning to mass-based systems.

Institutional and regulatory frameworks are pivotal to the smooth operation of an ETS. When developing their ETSs, economies of the Asia and Pacific region should dedicate significant resources to developing and refining the MRV systems that strengthen data management and accuracy. Legal mandates should be developed for facilities to report their GHG emissions data to ensure accurate emissions tracking and reporting, which will enhance system credibility and environmental integrity and lead to effective enforcement.

ETS institutional and regulatory frameworks are still being developed in many of the region's economies, as are the underpinning MRV systems. Policymakers charged with ETS portfolios need to provide clear mandates alongside roles and responsibilities to subsidiaries (e.g., state, provincial, and municipal governments) and other relevant bodies (e.g., registries, exchanges, third-party verifiers). To ensure smooth and effective ETS implementation, these stakeholders require predictable resources and staffing as well as substantial (and ongoing) capacity building and technical assistance from regional and international experts.

Reforming the energy sector to enable carbon price pass-through may be essential for an effective ETS. The energy sector in many economies in Asia and the Pacific is heavily regulated or in the early phases of liberalization. Regulated energy sectors hinder carbon price pass-through, negating the overall impact of an ETS. The PRC has for decades attempted varying levels of energy sector reform and, in 2020, it began a trial operation of eight pilot spot markets that utilize real-time electricity trading, formulate market-clearing prices, and guide electricity production and consumption through flexible market price signals. These spot markets have introduced price signals into electricity generation and distribution and, in doing so, they serve as an outlet for ETS cost pass-through to consumers. Their expansion, which will take time, along with further regulation and capacity building, could meaningfully improve the ability of the PRC's national ETS to reduce emissions. The PRC's experience shows that ETS architects and policymakers need to coordinate closely with power market reformers to promote a positive policy intersection, such as that created by the development of spot markets, enabling greater ETS effectiveness.

To add flexibility, countries can also allow for offset use in an ETS. However, such allowances should only be made to the extent that they do not undermine the effectiveness of the ETS. An ETS is inherently a flexible instrument because regulated entities can either purchase allowances from the market or reduce their own emissions, but allowing the use of offsets can provide additional flexibility. These offsets are generated through a crediting mechanism, which allows for the remuneration of achieved emissions reductions through the issuance of tradable certificates, while ensuring adherence to specific requirements. The crediting mechanism can be either domestic or international (section 5.5). Domestic offsets have been used in a range of ETS jurisdictions, e.g., subnationally in California and under the Regional Greenhouse Gas Initiative (encompassing Connecticut, Delaware, Maine, Maryland, Massachusetts, New Hampshire, New Jersey, New York, Pennsylvania, Rhode Island, and Vermont) and nationally in the PRC (Box 5.4). International offsets have historically played a greater role in Europe, the ROK, and New Zealand.

5.4.3   Emissions Trading System Linkages Across Economies

An international carbon pricing framework is needed to unify and achieve climate goals. Efforts to reduce emissions are fragmented due to divergent national climate ambitions and largely isolated ETSs. This highlights the need for an internationally coordinated carbon pricing system to achieve climate goals cost-effectively. The primary advantage of linking carbon markets is lowering emissions reduction costs (Anger 2008; Li and Duan 2021). Due to disparate marginal abatement costs, the cross-border exchange of carbon allowances can lower the costs of emissions compliance for all parties involved (Jaffe and Stavins 2008; Dellink et al. 2014). Economies exporting carbon permits benefit from sales revenue, while those importing permits benefit from reduced compliance costs (Qi and Weng 2016; Ranson and Stavins 2016). The basic concept of carbon market linkage is presented in Box 5.5.

Linked markets can enhance price discovery, liquidity provision, and risk management. Introducing more participants into carbon markets increases market liquidity and allows carbon prices to better reflect public information (Brunnermeier and Pedersen 2009; Santikarn, La Hoz Theuer, and Haug 2018; Brogaard, Hendershott, and Riordan 2019). This, in turn, contributes to the mitigation of price volatility (McKibbin, Shackleton, and Wilcoxen 1999; Jaffe and Stavins 2008; Flachsland, Marschinski, and Edenhofer 2009).

Box 5.4: A Domestic Voluntary Carbon Market—The China Certified Emission Reduction Scheme

Established in 2012, the China Certified Emission Reduction (CCER) scheme is central to the voluntary carbon market in the People's Republic of China (PRC) and plays a key role in the country's market-based carbon reduction strategy.

The CCER scheme features a domestic carbon credit offset trading standard used to monitor, report, and verify emission mitigation projects within the PRC. Carbon credits generated from a diverse range of projects—including renewable energy generation, methane capture, and forestry initiatives—have been traded in the CCER scheme, which provides avenues for businesses and individuals to voluntarily offset their emissions. Meanwhile, firms required to participate in the national compliance markets are allowed to purchase offset credits from the scheme at up to 5% of their quotation. Although the price level of CCER credits can sometimes fluctuate greatly, it is generally lower than that of the compliance market, so the scheme offers firms a flexible, often cost-effective way to achieve their emissions reduction targets.

Besides complementing the compliance market to achieve climate change mitigation, the CCER scheme helps channel substantial revenue from broad sources to targeted sectors such as renewable energy and forestry, which tend to be underinvested in a free market. For example, forestry projects under the CCER framework contribute to afforestation and reforestation, resulting in carbon sequestration and biodiversity conservation.

The CCER scheme also has the potential to become part of an international market, although it currently only operates within the PRC. It has largely adopted methodologies approved by the Clean Development Mechanism under the United Nations Framework Convention on Climate Change, and these methodologies are eligible for use with the Verified Carbon Standard program, the world's most widely used greenhouse gas crediting program.

The revamped CCER scheme, which was launched in early 2024, is the result of foresight by the Government of the PRC, which in 2017 suspended the issuance of new CCER credits in order to usher in a new era of effectiveness. The government recognized the need to streamline regulatory oversight, improve the quality of accredited projects, and reduce fraudulent claims and double-counting of emission reductions, through standardization and enhanced monitoring, reporting, and verification.

While keeping the strong features of the earlier scheme (e.g., its connection to the compliance market and use of internationally recognized methodologies), the new CCER scheme offers improved integrity, quality, and alignment with national climate policies. First, it expands the types of carbon offset projects allowed, enabling more clean energy projects to be registered and granted tradable CCER offset credits. Second, in line with the PRC's focus on integrating environmental and climate policies under a single authority, oversight of the new scheme has been shifted to the Ministry of Ecology and Environment, with this move expected to streamline regulatory protocols and enhance integration between the compliance and voluntary markets. Third, the new CCER scheme emphasizes transparency in reports for design, validation, verification, and data, which is expected to ensure consistency and stabilize prices.

These reforms to the CCER scheme are a significant milestone in the PRC's carbon reduction strategy, with the improvements attracting a wider range of participants and fostering greater confidence in both the domestic and international voluntary carbon markets.

Source: Gong, Y. 2024. The Voluntary Carbon Market, Globally and in the PRC: History, Current Situation and Prospects. Background paper for the *Asia–Pacific Climate Report 2024: Catalyzing Finance and Policy Solutions.* Asian Development Bank.

Box 5.5: Concept of Carbon Market Linkage

Carbon markets are considered linked if a participant in one market can use a carbon unit issued under another market to meet compliance obligations (Haites 2004; Haites 2016). Linkage between carbon markets can assume various forms, with differences in the degree, scope, and direction of trading flows, as shown in the figure below.

Types of Carbon Market Linkages

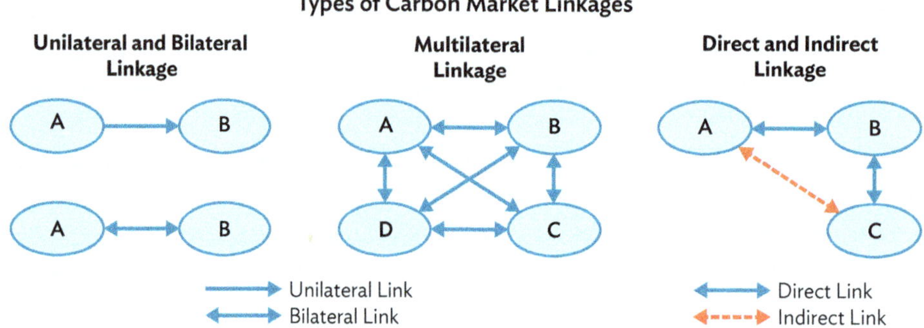

Source: Mehling, M. 2024. Theories and Practices of Cross-Border Carbon Market Linkages: Implications for Asia. Background paper for the *Asia–Pacific Climate Report 2024: Catalyzing Finance and Policy Solutions*. Asian Development Bank.

Direct links allow trade both within individual systems and between different systems (Ellis and Tirpak 2006). These links can be unilateral, bilateral, or multilateral. Bilateral or multilateral linkages offer the greatest economic benefits by increasing market size and diversity, enhancing integration efficiency, and leveling compliance costs through price convergence (Mehling and Haites 2009). A direct link can also be purely unilateral, which is appealing because it can be implemented without prior negotiations or mutual commitments.

An indirect linkage occurs when one carbon market links to a second system which, in turn, is linked to a third. Though not directly connected, the first and third markets influence each other through their mutual link to the second market, affecting unit supply, demand, and prices (Mehling and Haites 2009). Indirect links have commonly arisen when different markets accept the same offset credits, such as Certified Emission Reductions or Emission Reduction Units from the Kyoto Protocol mechanisms (Dellink et al. 2014; ADB 2016). Similarly, carbon credits authorized for international use under Article 6 of the Paris Agreement on Climate Change can be used to create indirect links.

References:
ADB (Asian Development Bank). 2016. *Emission Trading Schemes and Their Linking: Challenges and Opportunities in Asia and the Pacific*.
Dellink, R., S. Jamet, J. Chateau, and R. Duval. 2014. *Towards Global Carbon Pricing: Direct and Indirect Linking of Carbon Markets*. Organisation for Economic Co-operation and Development (OECD).
Ellis, J. and D. Tirpak. 2006. *Linking GHG Emission Trading Schemes and Markets*. COM/ENV/EPOC/IEA/SLT(2006) 6. OECD.
Haites, E. 2004. Tradeable Permits in the Policy Mix and Harmonisation of Emissions Trading Schemes. In OECD, ed. *Greenhouse Gas Emissions Trading and Project-Based Mechanisms*. OECD. pp. 103–18.
Haites, E. 2016. Experience with Linking Greenhouse Gas Emissions Trading Systems. *WIREs Energy and Environment*. 5 (3). Wiley Interdisciplinary Reviews. pp. 246–60.
Mehling, M. and E. Haites. 2009. Mechanisms for Linking Emissions Trading Schemes. *Climate Policy*. 9 (2). pp. 169–184.
Background paper for the *Asia–Pacific Climate Report 2024: Catalyzing Finance and Policy Solutions*. Asian Development Bank.

Source: Mehling, M. 2024. Theories and Practices of Cross-Border Carbon Market Linkages: Implications for Asia. Background paper for the *Asia–Pacific Climate Report 2024: Catalyzing Finance and Policy Solutions*. Asian Development Bank.

ETS linkage can reduce "carbon leakage." Due to the varying strictness of climate policies across different economies, regulated firms may consider relocating to jurisdictions with less-stringent regulations (a phenomenon known as carbon leakage), which undermines global efforts to reduce GHG emissions. Notably, the convergency of carbon prices in linked ETSs mitigates this risk (World Bank 2015).

Early unilateral carbon market linkages proved easy to establish and terminate. Carbon market linkages began in Europe and North America in the early 2000s. The Chicago Climate Exchange allowed Certified Emission Reductions in 2003 and allowances from the EU's ETS in 2005. However, these moves saw minimal uptake, leading to termination (Haites 2015). Norway linked its market unilaterally to the EU's ETS in 2005, fully integrating bilaterally in 2012. ETS markets in the EU, Japan, the ROK, New Zealand, Norway, and Switzerland linked project-based mechanisms under the Kyoto Protocol.[14] The Tokyo Metropolitan Government linked its ETS with Saitama prefecture's in 2011. These experiences showed unilateral linkages are easy to start and end, while linkage to international credits decreased due to concerns over additionality, ensuring that emissions reductions would not have occurred with the reduction activity and the environmental integrity of Clean Development Mechanism units.

Successful long-term linkages rely on harmonization, transparency, support, and trust. Two prominent examples demonstrate that long-term ETS linkages rely on mutual recognition, regulatory harmonization, transparency, and coordinated administrative support to maintain market integrity and stakeholder trust. As the first example, California and Québec linked their ETSs in 2014. Ontario joined in 2017, but withdrew in 2018 after a change of government (Raymond 2020). Despite Ontario's withdrawal, California and Québec continued joint auctions and market activities, emphasizing transparency and coordinated administrative support for effective market management. As the second example, the EU and Switzerland signed an agreement to link their ETSs on 23 November 2017, effective 1 January 2020. The agreement includes mutual recognition of allowances, technical standards, and protection of sensitive information. It also established a joint committee for administration and dispute resolution and allows potential linkages with third parties. Transactions between the EU and Swiss registries became operational in 2021, ensuring system integrity and transparency.

Differences in institutional settings, climate goals, and policy objectives add significant challenges. While procedural disparities—such as sectoral and GHG coverage and allowance allocation methods—can be harmonized, differences in legal frameworks and climate policy legislation can impede ETS consensus among economies (Fankhauser, Gennaioli, and Collins 2015). Varying ambitions on climate goals also further complicate linkages. Economic circumstances may lead governments to scale back emissions reduction initiatives and relax enforcement standards. Effective cross-border carbon markets require harmonization among ETSs, which can conflict with domestic policy objectives (Flachsland, Marschinski, and Edenhofer 2009). Maintaining control over individual carbon market policies also poses a challenge, as bilateral and multilateral linkages necessitate high consistency in cap settings, allowance allocation, and MRV processes.

A focus on domestic markets can provide a platform for future cross-border linkages. Given the complexity of establishing an extensive cross-border carbon market linkages, each compliance market should prioritize further development of domestic markets while aiming to leverage the benefits from carbon market connectedness. To facilitate the linking process, the following six policy recommendations are suggested: (i) establish a communication forum or leverage international organizations to share policies and coordinate efforts, ensuring transparency and consistent information flow; (ii) implement a legally-backed MRV framework with enforcement measures to ensure accurate emissions tracking and reporting, enhancing system credibility and environmental integrity; (iii) create an authorized cooperative organization and robust trading infrastructure for carbon-related products, including post-trade registries; (iv) ensure accurate measurement of international carbon allowances to mitigate double-issuing and double-using of carbon credits; (v) implement cooperative policies to maintain environmental integrity, promote common objectives such as nationally determined contributions, and incentivize future mitigation actions; and (vi) strengthen financial regulations and policies to mitigate risks—such as fraud, misleading information, manipulation, and transfer pricing—in carbon finance.

---

14   Since 2020, the EU no longer accepts the use of international credits, including Certified Emission Reductions, for compliance under its ETS.

5.5   The Role of International Carbon Markets

5.5.1   Recent Development of International Carbon Markets

There is an emerging landscape of international carbon markets for cost-effective emissions reduction. International carbon markets enable the trading of carbon credits internationally, both through the compliance market under Article 6 of the Paris Agreement and through voluntary carbon markets (VCMs). These markets are based on baseline-and-crediting mechanisms, whereby carbon credits are generated through investments that reduce emissions relative to a counterfactual "baseline" estimate of future emissions. Independent carbon crediting programs provide systems for the generation of carbon credits, as does Article 6.4, also known as the Paris Agreement Crediting Mechanism (PACM). These markets are important as they mobilize carbon finance, which can enhance the attractiveness of private investment in low-carbon technologies and solutions. Governments and private companies can therefore utilize international carbon markets to reduce emissions cost-effectively and access funding for sustainable development.

Article 6 lays the foundation for international compliance markets under the Paris Agreement. Article 6 enables both developing and developed economies to buy and sell Internationally Transferred Mitigation Outcomes (ITMOs) to achieve their mitigation commitments under nationally determined contributions (NDCs). Within Article 6, international cooperation on climate mitigation can be pursued through two market-based approaches established under Article 6.2 and Article 6.4, respectively. Article 6.2 is a decentralized approach that offers greater flexibility in defining the scope and methods of cooperation while using common accounting and reporting rules. For example, under Article 6.2, market scope and methods can be defined bilaterally between cooperating Parties while complying with the guidance for accounting and reporting from the United Nations Framework Convention on Climate Change (UNFCCC). Article 6.4, now termed the PACM, uses a centralized approach and is governed by a supervisory body of the UNFCCC. Since the adoption of the Paris Agreement, significant progress has been made in clarifying how these market mechanisms are intended to function. Under Article 6 rules, decisions have been taken that the trading of ITMOs under the Paris Agreement will require authorization from the host government and the application of a corresponding adjustment to avoid double-counting with host economy NDCs. However, additional decisions are needed to reduce uncertainty for Article 6.2 and to operationalize the PACM.

International carbon markets under Article 6 can generate significant global financing. While estimates of such financing fall within a wide range, studies suggest the potential economic benefits of Article 6 cooperation are significant. The World Bank estimates that resource flows generated through international carbon trading, under full and cost-effective implementation of NDC targets, need to be $185 billion in 2030 (World Bank 2016). The International Emissions Trading Association notes Article 6 market projections of $250 billion a year by 2030 and $1 trillion a year by 2050 (Edmonds et al. 2019).

While international markets are still developing, economies in Asia and the Pacific are already active. As of 5 July 2024, there were 83 bilateral carbon trading agreements between 10 different buyers and 46 host economies globally, with 140 pilot projects recorded (of which 119 belong to Japan's Joint Crediting Mechanism) (UNEP 2024). In Asia and the Pacific, Thailand and Vanuatu have provided national authorization for the export of credits under Article 6 and the first-ever issuance of 1,916 ITMOs occurred for the E-Bus Program in Bangkok in 2023: this transaction between Thailand and Switzerland is the first under Article 6.2 (UNEP 2024).

The most immediate opportunity for issuance of carbon credits through the PACM relates to the transition of Clean Development Mechanism (CDM) activities. Parties to the Paris Agreement agreed that projects meeting eligibility requirements should be able to transition from the CDM to the PACM and thus continue to generate carbon credits. According to an assessment by the United Nations Environment

Programme, as of 5 July 2024, transition requests were correctly submitted for 1,356 project activities and 118 programs of activities (including 949 component project activities). Total potential emissions reductions for 2021–2025 generated from transition CDM activities could amount to more than 900 million metric tons of CO_2 as per the established CDM methodologies (UNEP 2024).

Climate-conscious enterprises can benefit greatly from voluntary carbon markets. International VCMs can be utilized by organizations, typically in the private sector, to support achieving their own voluntary mitigation commitments, such as pledges to achieve carbon neutrality or net zero emissions. International VCMs encompass a collection of standard bodies, voluntary buyers, and carbon credit generators that sell verified carbon credits to be used for voluntary purposes. Independent carbon crediting programs or standards, such as the Gold Standard and VERRA's Verified Carbon Standard, set the requirements for delivering and measuring emissions reductions and removals resulting from a mitigation activity.

VCMs must rebound from falls in trading values and volumes in 2022 and 2023. The total trading value of all VCMs (international) peaked in 2021, reaching about $2.1 billion, then declined to $1.9 billion in 2022 and to $0.72 billion in 2023. Transaction volumes of VCM-traded carbon credits in metric tons of carbon dioxide equivalent (tCO$_2$e) also peaked at 516 million in 2021, then fell to 254 million in 2022 and to 111 million in 2023. The reported average nominal price in 2023 was $6.53 per tCO$_2$e, an 11% decline on 2022, but over 60% above the reported average nominal price in 2021 (Procton 2024). There is a mix of concerns regarding VCMs, particularly their high transactions costs and demand uncertainty. There are also issues regarding overall market integrity related to the varying quality of credits, due to inflated baselines and challenges of monitoring, as well as negative impacts on local communities. However, despite trading declines and obstacles hindering growth, the overall sentiment for VCMs is optimistic.

Efforts are being made to address integrity issues in VCMs. Two notable initiatives for voluntary markets are the Integrity Council for the Voluntary Carbon Market (ICVCM) and the Voluntary Carbon Markets Integrity Initiative (VCMI). These two bodies are establishing guidelines and standards for high-integrity VCMs on the supply side and demand side, respectively. The publication of the ICVCM's Core Carbon Principles and the launch of the VCMI's Claims Code are contributing to buyer confidence in market quality and integrity. In addition to embracing ICVCM and VCMI guidelines, governments are taking their own measures to build confidence in VCMs. For example, the Government of the United States has announced new principles to advance the responsible development of VCMs, thereby encouraging these markets to embrace identified approaches for better supply, demand, and market-level integrity. Papua New Guinea, meanwhile, is regulating nature-based solutions projects in particular, with a benefit-sharing mechanism as well as stringent, free, prior, and informed content provisions.

Industry-specific approaches can complement broader international carbon schemes. Alongside Article 6 and the VCMs, there are industry-specific approaches that influence international carbon markets. One example is the Carbon Offsetting and Reduction Scheme for International Aviation. International aviation is not included in national emissions inventories nor in NDCs, and has instead been addressed by the International Civil Aviation Organization (ICAO). In 2016, the ICAO Assembly adopted a global market-based emissions trading scheme for international aviation. A technical committee under the ICAO has determined eligibility criteria for carbon credits, allowing independent carbon crediting programs, the PACM, and domestic carbon crediting programs to generate carbon credits. The credits will need authorization under Article 6.2 of the Paris Agreement (Anjaparidze 2019).

5.5.2   Opportunities and Challenges for Asia and the Pacific

Developing economies can use international carbon markets to meet domestic climate goals. With the finalization of the Paris Agreement Rulebook, many countries are operationalizing international carbon markets under Article 6 as well as engaging with VCMs. Although developing economies in Asia and the Pacific are mostly exploring avenues to export carbon credits, there is also potential for them to meet their climate ambitions by buying credits through international carbon markets. This could occur when an economy has projected that it may not reach its NDC targets (especially when these targets have been recently lifted) or when purchasing credits proves to be cheaper than domestic mitigation efforts (which may have high marginal abatement costs). Assuming that using public funds to purchase carbon credits is politically challenging, this task can be delegated to the private sector. For instance, economies with a high domestic carbon price may decide to tighten the cap of their ETS but allow for private entities under the cap to purchase international carbon credits (ITMOs), since this may constitute a more cost-effective alternative to high ETS prices. This has been the case in Indonesia.

The Asia and Pacific region has extensive experience in carbon market mechanisms. Approximately 80% of all CDM projects are based in Asia and the Pacific (Amarjargal et al. 2020). While project developers and validation and verification bodies can draw on experiences from the CDM, host economies have less to build on since Article 6 requires greater host involvement and higher institutional and MRV requirements. Japan's Joint Crediting Mechanism (JCM) is a forerunner to Article 6.2 and around 90% of all JCM projects are hosted in the region (IGES 2024). Developing economies in Asia and the Pacific can draw on experiences from the JCM as a bilateral approach.

The region has work to do in scaling up involvement in international carbon markets. Some of the existing challenges include: (i) lack of national strategies and frameworks for engagement with international carbon markets; (ii) ineffective governance systems including lack of interministerial coordination as well as lack of mandates and legal instruments; (iii) lack of institutional capacity and/or skills of personnel to negotiate bilateral agreements; and (iv) lack of the carbon market expertise required to manage domestic processes and procedures for engagement in international markets.

National carbon market strategies are a vital catalyst for international engagement. Alongside enhancing institutional capacity and technical readiness to participate in international carbon markets, an important first step for many economies is to develop holistic national carbon market strategies (ADB 2023b). To engage strategically in carbon markets, policymakers need to weigh the benefits of selling mitigation outcomes to attract carbon finance against the need to meet their NDC targets. Evaluating this trade-off requires a long-term strategy that aims to achieve NDC targets and informs engagement with Article 6, including guiding principles and criteria for mitigation outcomes to be eligible for international transfer. Such a strategy must embrace the interrelationship between various carbon markets and cover management of both the compliance and voluntary markets. For instance, buyers in a VCM may wish to have carbon credits (ITMOs) that have authorization and cannot be used toward a host economy's NDC targets. Another type of integration can occur when domestic carbon crediting mechanisms are used for Article 6.2 transactions, which may require new regulations and legal frameworks.

The NDC accounting aspect of Article 6 is critical to effective market integration. For economies to engage with international carbon markets, they need to have a framework outlining the sectors and types of mitigation activities from which they will allow ITMO exports. This, in turns, requires an understanding of how the NDC system will be implemented. The framework also needs to describe administrative procedures to manage ITMO authorization, tracking of ITMOs, and their accounting (corresponding adjustments). Governments may consider what actions need to be taken to promote voluntary market activities within their jurisdiction and may consider regulating such markets.

5.5.3   Support for Nature-Based Solutions Through International Carbon Markets

Nature-based solutions (NBSs) have the potential to ramp up global climate finance. As global efforts intensify to achieve net zero emissions by 2050, integrating NBSs into international carbon markets offers a pathway to mobilize financial resources at scale. These solutions leverage the power of nature and healthy ecosystems to protect people, optimize infrastructure, and safeguard a stable and biodiverse future (Cohen-Shacham et al. 2016). They include activities such as climate-smart agriculture, biodiverse habitat restoration, water resource management, clean cookstove projects, and green infrastructure initiatives. A significant aspect of NBSs is their long-term benefits. For instance, sustainable land-use change can mean soils continue to store CO_2 even after global emissions peak, reducing temperatures in the long term and contributing to planetary cooling. In fact, NBSs have the potential to provide up to 37% of the mitigation necessary by 2030 to reach the targets of the Paris Agreement (Griscom et al. 2017).

Despite these benefits, NBSs face a significant financing gap. Required investment in NBSs stood at $200 billion annually in 2022 and needs to triple by 2030 and quadruple by 2050 to reach climate, biodiversity, and land degradation targets (UNEP 2023). The primary challenge lies in the undervaluation of nature's benefits within traditional economic frameworks and conventional financing systems, leading to insufficient private investment. Higher risks, lower returns, and longer payback periods discourage investment in NBSs, despite their potential to enhance climate and ecosystem resilience (UNEP 2023). Public finance therefore has been the primary source of finance flows to NBSs, accounting for 82% in 2023, with a focus on biodiversity, landscape protection, sustainable agriculture, forestry, and fishing. Private finance accounts for only 18% of total finance flowing to NBSs, with more than half channeled through biodiversity credits, offsets, and sustainable supply chains (UNEP 2023).

Carbon markets present an opportunity to address this gap. They can do so by assigning a financial value to carbon sequestration and emission reductions through nature. It is estimated that NBSs financed through carbon markets can deliver 10%–12% of the mitigation needed by 2030 to be on track to reach global net zero by 2050 (Landholm et al. 2022). These estimations account for price, implementation feasibility, and spatial location constraints.

Agriculture, forestry, and wetlands management can be driving forces in carbon markets. The core potential of NBSs in carbon markets is projected to be in the agriculture sector, forestry (avoided deforestation,[15] afforestation and/or reforestation, and improved forest management), and the restoration and conservation of wetlands. Using 2030 as the benchmark, agriculture at 43% has the highest carbon market potential under NBS measures, followed by avoided deforestation at 32%, afforestation and/or reforestation at 11%, improved forest management at 7%, and the restoration and conservation of wetlands at 7% (Landholm et al. 2022). The top five economies with the highest potential in agriculture carbon markets are the United States, the PRC, India, Brazil, and Indonesia. The potential for avoided deforestation is largest in Latin America and Asia, particularly in Brazil, Indonesia, Colombia, Peru, and Malaysia. Considering land mass as the measure, the largest mitigation potential across all NBS carbon market activities lies in Indonesia, Malaysia, Costa Rica, Colombia, and Belize (Landholm et al. 2022).

---

[15]   Negotiations at the 60th session of the Subsidiary Body for Scientific and Technological Advice (SBSTA 60) in Bonn, 3–13 June 2024, included draft decisions to refrain from carbon crediting of avoided deforestation emissions until 2028, when the issue will be discussed again. This implies that avoided deforestation as a mitigation activity is only eligible on the VCMs.

Prices for NBS carbon credits vary depending on the solution type, project, and market used. However, they typically range from $5 to $15 per t$CO_2$e, reflecting the market's recognition of their interrelated environmental and social benefits (Procton 2024). As global climate ambitions grow, demand for high-quality NBS credits is expected to rise, creating new opportunities for investment and innovation.

The future of NBSs in international carbon markets appears complex yet promising. Despite its high potential, the NBS market must overcome several challenges, including the need for robust MRV systems to ensure the integrity and transparency of the carbon credits and for protocols that can underpin the permanence of NBS efforts. Policy frameworks must evolve to support the scaling up of NBS projects. This includes addressing barriers such as land tenure issues, regulatory complexities, and access to high-integrity carbon finance for local communities. Nevertheless, integrating NBSs into international carbon markets offers a promising means by which to mobilize resources for climate mitigation and sustainable development. By valuing ecosystem services and carbon sequestration through market mechanisms, carbon markets can incentivize investment in NBS projects that deliver environmental, social, and economic benefits.

5.6   The Effects of Carbon Pricing on Developing Economies

Despite strong momentum toward carbon pricing, there are some concerns for developing economies. In particular, there has been speculation that efforts to implement carbon pricing may lead to a loss of global competitiveness. An additional concern is the risk of carbon leakage in response to carbon pricing, with firms shifting production out of certain jurisdictions to avoid paying a carbon price, thereby reducing the ability of governments to regulate GHG emissions.

The impact of carbon pricing on competitiveness is ambiguous. While the imposition of carbon pricing will increase costs for local firms that are subjected to the carbon price, it can also be an incentive to modernize production techniques to drive improvements in productivity and competitiveness. Consistent with the Porter Hypothesis, a carbon price can also encourage innovation in new green technologies, leading to improvements in competitiveness by avoiding the burden of carbon pricing and increasing both demand (through consumer preference for climate-friendly brands) and profits. Empirical evidence suggests that the competitiveness effects of carbon pricing are usually quite small (see, for example, Arlinghaus 2015). However, studies of existing carbon pricing schemes have been undertaken in the context of low carbon prices and may not reflect effects of carbon pricing at higher levels.

The overall impact also depends on the global coverage of carbon pricing schemes. With carbon pricing not occurring globally, there is the risk that firms operating in economies with carbon pricing may lose market share and see reduced profits, with production shifting to jurisdictions with no carbon pricing (or where carbon is priced lower). The extent and effect of this carbon leakage will be diminished with increased coordination and coverage of carbon pricing schemes across the world, because it will become more difficult for firms to shift production beyond carbon pricing boundaries.

More is revealed about competitiveness impacts when advanced modeling is applied. Adopting a computable general equilibrium model allows for an examination of the competitiveness impacts of carbon pricing under different pricing and coverage scenarios. The modeling exercise starts with a scenario that examines the impact of the implementation of the Carbon Border Adjustment Mechanism (CBAM) by the EU (Table 5.2). The CBAM is intended to provide surety for EU firms by charging an equivalent carbon price on imported intermediate inputs as on domestic inputs. Given the potential impact of this initiative on the competitiveness of economies in Asia and the Pacific, the initial scenario assumes the imposition of the CBAM and an increase in the EU's ETS carbon price to €100 per metric ton of CO_2, from a baseline of €18.[16] With other developed economies likely to follow the EU in implementing Border Carbon Adjustment (BCA) mechanisms, a second scenario involves extending the CBAM to all Organisation for Economic Co-operation and Development (OECD) economies (including those in Asia and the Pacific). The remaining scenarios implement carbon pricing, including BCA, in specific subregions of Asia and the Pacific to examine the estimated competitiveness effects of carbon pricing on these subregions, allowing for substitution effects toward subregions without carbon pricing. To account for the additional challenges for least-developed economies, additional scenarios are introduced that involve these economies charging a lower carbon price of €50. The final scenario introduces carbon pricing across all of Asia and the Pacific, with the main substitution effects then being toward other regions of the world.

Table 5.2: Overview of Carbon Pricing Scenarios

Scenario 1	- European economies impose tighter ETS carbon allocations, with a resulting €100/MT price. - CBAM taxes are imposed for ETS sectors.
Scenario 2	- All OECD economies impose tighter ETS carbon allocations, with a resulting €100/MT price. - CBAM taxes are imposed for ETS sectors.
Scenario 3	- OECD economies impose tighter ETS carbon allocations, with a resulting €100/MT price. - Non-LDE SE Asian countries impose tighter ETS carbon allocations, with a resulting €100/MT price. - CBAM taxes are imposed for ETS sectors.
Scenario 4	- OECD economies impose tighter ETS carbon allocations, with a resulting €100/MT price. - Non-LDE East Asian economies impose tighter ETS carbon allocations, with a resulting €100/MT price. - CBAM taxes are imposed for ETS sectors.
Scenario 5	- OECD economies impose tighter ETS carbon allocations, with a resulting €100/MT price. - Non-LDE South Asian countries impose tighter ETS carbon allocations, with a resulting €100/MT price. - CBAM taxes are imposed for ETS sectors.
Scenario 6	- OECD economies impose tighter ETS carbon allocations, with a resulting €100/MT price. - Non-LDE Other Asian countries impose tighter ETS carbon allocations, with a resulting €100/MT price. - CBAM taxes are imposed for ETS sectors.
Scenario 7	- OECD economies impose tighter ETS carbon allocations, with a resulting €100/MT price. - All non-LDE Asian countries impose tighter ETS carbon allocations, with a resulting €100/MT price. - CBAM taxes are imposed for ETS sectors.
Scenario 8	- OECD economies impose tighter ETS carbon allocations, with a resulting €100/MT price. - All non-LDE Asian countries impose tighter ETS carbon allocations, with a resulting €100/MT price. - All LDE Asian countries impose tighter ETS carbon allocations, with a resulting €50/MT price. - CBAM taxes are imposed for ETS sectors.

CBAM = Carbon Border Adjustment Mechanism, ETS = emissions trading system, LDE = least-developed economy, MT = metric ton, OECD = Organisation for Economic Co-operation and Development.
Notes: During the phase-in period, the CBAM regime will not apply to all ETS sectors. However, the CBAM system is expected to be expanded to all ETS sectors after the phase-in period. Central and West Asia comprises Armenia, Azerbaijan, Georgia, Kazakhstan, the Kyrgyz Republic, Pakistan, Tajikistan, Turkmenistan, and Uzbekistan. LDE South Asia comprises Bangladesh and Nepal. LDE Southeast Asia comprises Cambodia, the Lao People's Democratic Republic, Myanmar, and Timor-Leste. Other South Asia comprises Bhutan, India, Maldives, and Sri Lanka. Other Southeast Asia comprises Brunei Darussalam, Indonesia, Malaysia, the Philippines, Singapore, Thailand, and Viet Nam.
Source: Asian Development Bank.

---

[16]   The baseline of €18 per metric ton of CO_2 reflects the approximate price of CO_2 in the reference year of 2017. The revenue collected from the CBAM is assumed to go into the EU's budget in the model.

Imposing the CBAM and increasing the EU carbon price lowers estimated global output levels and exports. This outcome is consistent with strong income effects, with the higher costs of production in the EU leading to lower global output. The EU is most strongly affected, with gross domestic product (GDP) estimated to fall by almost 2%. Effects in other regions are generally muted, with those regions closer to the EU (e.g., Other Europe, West and North Africa, and Central and West Asia) tending to see somewhat larger declines compared to other regions (Figure 5.4). Even in these cases, however, reductions in GDP are estimated at less than 0.5%. Impacts on exports tend to be greater, with reductions of more than 1% in the EU, Other Europe, North and West Africa, Sub-Saharan Africa, Latin America, and Central and West Asia.

Figure 5.4: Modeled Impacts of the Carbon Border Adjustment Mechanism and Changes to the European Union's Emissions Trading System

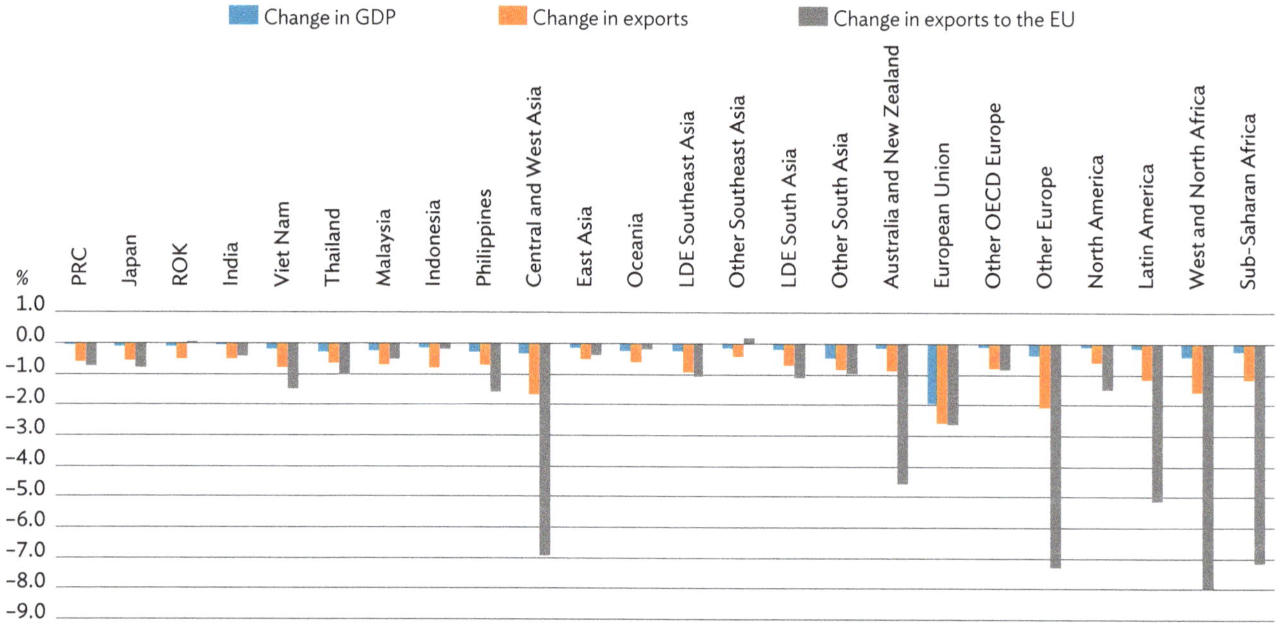

PRC = People's Republic of China, EU = European Union, GDP = gross domestic product, LDE = least-developed economy, OECD = Organisation for Economic Co-operation and Development, ROK = Republic of Korea.
Notes: Central and West Asia comprises Armenia, Azerbaijan, Georgia, Kazakhstan, the Kyrgyz Republic, Pakistan, Tajikistan, Turkmenistan, and Uzbekistan. LDE South Asia comprises Bangladesh and Nepal. LDE Southeast Asia comprises Cambodia, the Lao People's Democratic Republic, Myanmar, and Timor-Leste. Other South Asia comprises Bhutan, India, Maldives, and Sri Lanka. Other Southeast Asia comprises Brunei Darussalam, Indonesia, Malaysia, the Philippines, Singapore, Thailand, and Viet Nam.
Source: ADB calculations using data from the Global Trade Analysis Project 11 and the International Energy Agency (both accessed 15 November 2023).

Extending the ETS and CBAM to the rest of the OECD has varied effects. Many economies in Asia and the Pacific have estimated increases in output (Figure 5.5). While negative effects on output and exports within the EU diminish, effects in other OECD regions (Japan, the ROK, Other OECD Europe, and North America) become larger, with reductions in output and exports being relatively large in Japan and the ROK. In many economies and subregions of Asia and the Pacific, however, output levels (and in some cases exports) are estimated to rise. This is especially true for Other South Asia, Central and West Asia, Other Southeast Asia, the Philippines, Indonesia, and Thailand. These changes likely reflect carbon leakage, with downstream production shifting out of the OECD and into these economies.

Figure 5.5: Modeled Impacts of the Carbon Border Adjustment Mechanism and Emissions Trading Across OECD Economies

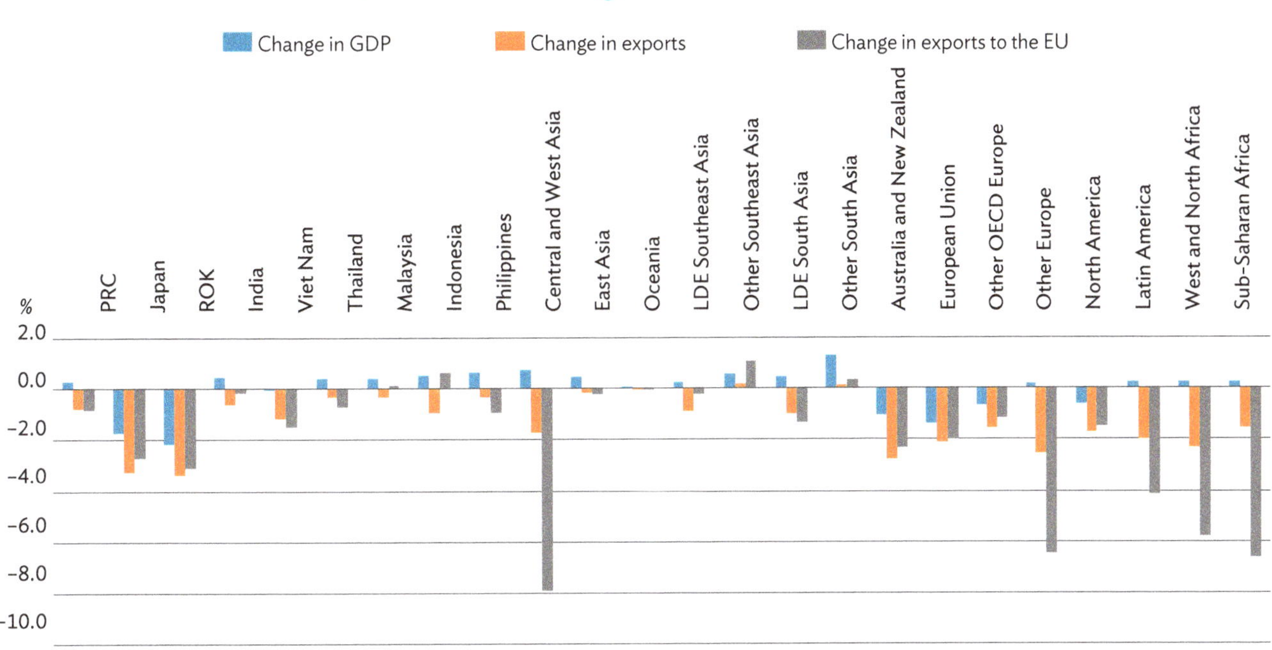

PRC = People's Republic of China, EU = European Union, GDP = gross domestic product, LDE = least-developed economy, OECD = Organisation for Economic Co-operation and Development, ROK = Republic of Korea.
Notes: Central and West Asia comprises Armenia, Azerbaijan, Georgia, Kazakhstan, the Kyrgyz Republic, Pakistan, Tajikistan, Turkmenistan, and Uzbekistan. LDE South Asia comprises Bangladesh and Nepal. LDE Southeast Asia comprises Cambodia, the Lao People's Democratic Republic, Myanmar, and Timor-Leste. Other South Asia comprises Bhutan, India, Maldives, and Sri Lanka. Other Southeast Asia comprises Brunei Darussalam, Indonesia, Malaysia, the Philippines, Singapore, Thailand, and Viet Nam.
Source: ADB calculations using data from the Global Trade Analysis Project 11 and the International Energy Agency (both accessed 15 November 2023).

Imposing carbon pricing selectively in subregions negatively impacts estimated output in those subregions. When a particular subregion imposes carbon pricing, it suffers reductions in output. In the case of East Asia, for example, output is estimated to fall by more than 2.2% in the PRC and 1.2% in the rest of East Asia (Table 5.3).[17] There is also strong evidence of carbon leakage in response to carbon pricing in subregions. Considering the case of East Asia, the imposition of carbon pricing is estimated to lead to substantial gains in GDP in other subregions and certain economies, with estimated increases of 4.6% in Other South Asia, 2.9% in Central and West Asia, 2.4% in the Philippines, 2.2% in Thailand, and 2.0% in Other Southeast Asia. Similar patterns are observed when looking at carbon pricing in other subregions.

Extending carbon pricing to all of Asia and the Pacific encourages carbon leakage and creates economic inequities. A coordinated policy of carbon pricing across the Asia and Pacific region (excluding least-developed economies) is estimated to have two main effects. First, it encourages carbon leakage to other regions, with output in West and North Africa estimated to increase by 2.3%, and with relatively large increases in Latin America (1.4%), Other Europe (1.8%), Sub-Saharan Africa (1.4%), and North America (0.8%). Second, it creates inequities within Asia and the Pacific. Reductions in output are estimated for East Asia and Oceania (the Pacific) and in the PRC, India, Malaysia, Thailand, and Viet Nam. Meanwhile, Indonesia, the Philippines, Central and West Asia, Other and Low-Income Southeast Asia, and Other and Low-Income South Asia are estimated to see output increase. This will partially reflect differences in CO_2 intensities, with the carbon price shifting production to those with relatively low intensities. Similar patterns are found to hold when introducing a (lower-priced) ETS and the CBAM in least-developed economies of Asia. In this scenario, these economies are still estimated to see an increase in output from region-wide carbon pricing, though the extent of this

---

[17]　The approach assumes that both the EU and the OECD also impose an ETS and the CBAM, meaning that the reported effects also include the effects of carbon pricing in these regions.

increase is diminished by its own carbon price. Such results lead to an important conclusion: even uniform carbon pricing in the Asia and Pacific region can have distributional consequences, depending on the carbon intensity and structure of production of the different economies and subregions.

Table 5.3: Estimated Output Changes in Response to Carbon Pricing Throughout Asia and the Pacific
(%)

	All OECD and Southeast Asia	All OECD and East Asia	All OECD and South Asia	All OECD and Other Asia and the Pacific	All OECD and All Non-LDE Asia	All OECD and All Asia
PRC	0.34	–2.15	0.47	0.32	–1.90	–1.89
Japan	–1.66	–0.33	–1.38	–1.66	0.17	0.18
Republic of Korea	–2.08	–1.06	–1.83	–2.08	–0.62	–0.60
India	0.51	1.33	–2.93	0.49	–1.92	–1.94
Viet Nam	–2.06	0.47	0.13	–0.01	–1.38	–1.32
Thailand	–4.87	2.20	0.81	0.44	–2.65	–2.45
Malaysia	–2.56	1.76	0.69	0.42	–0.85	–0.73
Indonesia	–0.52	1.90	0.83	0.56	1.21	1.30
Philippines	–1.56	2.37	1.11	0.73	0.72	0.80
Central and West Asia	0.80	2.87	1.23	–1.85	0.93	0.89
East Asia	0.51	–1.15	0.77	0.53	–0.72	–0.70
Oceania	–0.09	0.74	0.16	–2.56	–1.84	–1.80
LDE Southeast Asia	1.42	1.25	0.49	0.29	2.73	1.53
Other Southeast Asia	–1.46	1.96	0.86	0.58	0.20	0.38
LDE South Asia	0.47	1.68	0.61	0.49	1.97	1.00
Other South Asia	1.30	4.61	–3.24	1.46	0.29	0.22
Australia and New Zealand	–1.07	–0.66	–0.97	–1.03	–0.59	–0.58

PRC = People's Republic of China, LDE = least-developed economy, OECD = Organisation for Economic Co-operation and Development.
Notes: Central and West Asia comprises Armenia, Azerbaijan, Georgia, Kazakhstan, the Kyrgyz Republic, Pakistan, Tajikistan, Turkmenistan, and Uzbekistan. LDE South Asia comprises Bangladesh, Nepal. LDE Southeast Asia comprises Cambodia, the Lao People's Democratic Republic, Myanmar, Timor-Leste. Other Southeast Asia comprises Brunei Darussalam, Indonesia, Malaysia, the Philippines, Singapore, Thailand, and Viet Nam. Other South Asia comprises Bhutan, India, Maldives, and Sri Lanka.
Source: ADB calculations using data from the Global Trade Analysis Project 11 and the International Energy Agency (both accessed 15 November 2023).

Estimated impacts of carbon pricing on exports are largely consistent with those for output.
The estimated impacts on those economies and subregions implementing carbon pricing are generally in line with those for output, with exports declining substantially (Table 5.4). However, for the other economies and subregions not subject to carbon pricing, there are often differences between the output and export effects. In the case of carbon pricing by East Asia, for example, the estimates suggest increases in output but reductions in exports for Central and West Asia, least-developed economies in Southeast Asia and South Asia, and in India, Indonesia, and Viet Nam. Such results highlight that, while carbon pricing in one subregion can lead to increased output in other subregions, this does not necessarily equate to an increase in exports and global competitiveness. This further suggests that, in these cases, the increased output is linked to serving a higher share of domestic demand.

Table 5.4: Estimated Changes in Exports in Response to Carbon Pricing Throughout Asia and the Pacific
(%)

	All OECD and Southeast Asia	All OECD and East Asia	All OECD and South Asia	All OECD and Other Asia and the Pacific	All OECD and All Non-LDE Asia	All OECD and All Asia
PRC	−0.89	−2.09	−0.52	−0.77	−1.90	−1.89
Japan	−3.39	−2.25	−2.84	−3.18	−1.95	−1.95
Republic of Korea	−3.41	−2.66	−2.92	−3.30	−2.23	−2.24
India	−0.74	−0.21	−3.92	−0.55	−3.67	−4.07
Viet Nam	−2.32	−1.04	−1.01	−1.20	−2.16	−1.93
Thailand	−4.16	1.35	0.01	−0.40	−2.18	−2.00
Malaysia	−3.02	0.70	−0.13	−0.41	−1.81	−1.50
Indonesia	−2.26	−0.37	−1.27	−1.17	−2.33	−1.76
Philippines	−2.47	0.41	0.07	−0.30	−1.40	−1.14
Central and West Asia	−1.85	−1.37	−1.66	−2.42	−2.15	−3.01
East Asia	−0.21	−2.79	0.20	−0.08	−2.38	−2.39
Oceania	−0.25	1.10	0.16	−4.09	−3.02	−3.06
LDE Southeast Asia	0.46	−0.13	−0.75	−0.84	1.36	0.26
Other Southeast Asia	−2.78	1.53	0.32	−0.03	−1.44	−0.95
LDE South Asia	−1.13	−1.21	−1.85	−1.02	−2.22	−0.20
Other South Asia	−0.06	2.24	−2.44	0.23	−0.44	−0.70
Australia and New Zealand	−3.08	−3.41	−2.91	−2.80	−3.91	−3.90

PRC = People's Republic of China, LDE = least-developed economy, OECD = Organisation for Economic Co-operation and Development.
Notes: Central and West Asia comprises Armenia, Azerbaijan, Georgia, Kazakhstan, the Kyrgyz Republic, Pakistan, Tajikistan, Turkmenistan, and Uzbekistan. LDE South Asia comprises Bangladesh, Nepal. LDE Southeast Asia comprises Cambodia, the Lao People's Democratic Republic, Myanmar, Timor-Leste. Other Southeast Asia comprises Brunei Darussalam, Indonesia, Malaysia, the Philippines, Singapore, Thailand, and Viet Nam. Other South Asia comprises Bhutan, India, Maldives, and Sri Lanka.
Source: ADB calculations using data from the Global Trade Analysis Project 11 and the International Energy Agency (both accessed 15 November 2023).

Potential effects of carbon pricing call for greater coordination and new compensation mechanisms.
The results of the modeling exercise show that imposing carbon prices unilaterally can have negative consequences for output and exports. Many of these effects disappear with increased coordination and coverage of carbon pricing, though distributional effects still result in economic inequities for economies and subregions. Such results highlight the need for increased efforts at global cooperation in encouraging the widespread adoption of carbon pricing schemes. Economies can provide tax rebates proportional to production in emitting industries to support competitiveness, along with applying border carbon adjustments to tax carbon-intensive imports. Yet, they cannot avoid distortions and leakage unless carbon pricing schemes are extended to a broader set of countries worldwide.

References

ADB (Asian Development Bank). 2016. *Emission Trading Schemes and Their Linking: Challenges and Opportunities in Asia and the Pacific.*

———. 2023a. *Asia in the Global Transition to Net Zero: Asian Development Outlook 2023 Thematic Report.*

———. 2023b. *National Strategies for Carbon Markets Under the Paris Agreement—Making Informed Policy Choices.*

———. 2024. *2021 Energy Policy of the Asian Development Bank Supporting Low-carbon Transition in Asia and the Pacific.*

Amarjargal, B., H. Ebro, J. Nylander, and V. K. Duggal. 2020. *Achieving Nationally Determined Contributions through Market Mechanisms in Asia and the Pacific. Asian Development Bank Sustainable Development Working Paper Series.* No. 64.

Anger, N. 2008. Emissions Trading Beyond Europe: Linking Schemes in a Post-Kyoto World. *Energy Economics.* 30 (4). pp. 2028–2049.

Anjaparidze, G. 2019. The Extraordinary Climate Agreement on International Aviation: An Airline Industry Perspective. Policy Brief. Harvard Project on Climate Agreements. October.

APEC (Asia-Pacific Economic Cooperation). 2021. Aotearoa Plan of Action (Annex to 2021 APEC Economic Leader's Declaration).

Ari, A., N. Arregui, S. Black, O. Celasun, D. M. Iakova, A. Mineshima, V. Mylonas, et al. 2022. Surging Energy Prices in Europe in the Aftermath of the War: How to Support the Vulnerable and Speed up the Transition Away from Fossil Fuels. *IMF Working Paper.* No. 2022/152. International Monetary Fund.

Arlinghaus, J. 2015. Impacts of Carbon Prices on Indicators of Competitiveness: A Review of Empirical Findings. *OECD Environment Working Papers.* No. 87. Organisation for Economic Co-operation and Development.

Beaton, C., I. Gerasimchuk, T. Laan, K. Lang, D. Vis-Dunbar, and P. Wooders. 2013. *A Guidebook to Fossil-Fuel Subsidy Reform for Policy-Makers in Southeast Asia.* International Institute for Sustainable Development.

Beaton, C. 2023. *Government Support for Carbon-Intensive Energy in Asia: Policy Actions.* ADB.

Bernard, J., M. Kichian, and M. Islam. 2018. Effects of BC's Carbon Tax on GDP. *USAEE Research Paper Series.* No. 18. 329. United States Association for Energy Economics.

Best, R., P. J. Burke, and F. Jotzo. 2020. Carbon Pricing Efficacy: Cross-Country Evidence. *Environmental Resource Economics.* 77. pp. 69–94. 19 June.

Bistline, J., N. Mehrotra, and C. Wolfram. 2023. Economic Implications of the Climate Provisions of the Inflation Reduction Act. *NBER Working Paper Series.* May. 31267.

Bridle, R., S. Sharma, M. Mostafa, and A. Geddes. 2019. *Fossil Fuel to Clean Energy Subsidy Swaps: How to Pay for an Energy Revolution.* International Institute for Sustainable Development.

Brogaard, J., T. Hendershott, and R. Riordan. 2019. Price Discovery without Trading: Evidence from Limit Orders. *The Journal of Finance.* 74 (4). pp. 1621–58.

Brunnermeier, M. and L. H. Pedersen. 2009. Market Liquidity and Funding Liquidity. *Review of Financial Studies.* 22 (6). pp. 2201–2038.

Carbon Pricing Leadership Coalition. 2019. *Report of the High-Level Commission on Carbon Pricing and Competitiveness.* World Bank.

Coady, D., C. Flamini, and L. Sears. 2015. *The Unequal Benefits of Fuel Subsidies Revisited: Evidence for Developing Countries.* International Monetary Fund.

Coady, D., S. Fabrizio, B. Shang, A. Kangur, M. Nozaki, I. Parry, V. Thakoor, et al. 2013. *Case Studies on Energy Subsidy Reform: Lessons and Implications.* International Monetary Fund.

Cohen-Shacham, E., C. Janzen, S. Maginnis, and G. Walters. 2016. *Nature-Based Solutions to Address Global Societal Challenges.* International Union for Conservation of Nature (IUCN).

Dellink, R., S. Jamet, J. Chateau, and R. Duval. 2014. *Towards Global Carbon Pricing: Direct and Indirect Linking of Carbon Markets.* Organisation for Economic Co-operation and Development.

Dhakal, S., J. C. Minx, F. L. Toth, A. Abdel-Aziz, M. J. Figueroa Meza, K. Hubacek, I. G. C. Jonckheere, et al. 2022. Emissions Trends and Drivers. In P. R. Shukla, et al., eds. *Climate Change 2022: Mitigation of Climate Change. Contribution of Working Group III to the Sixth Assessment Report of the Intergovernmental Panel on Climate Change.* Cambridge University Press.

Döbbeling-Hildebrandt, N., K. Miersch, T. M. Khanna, M. Bachelet, S. B. Bruns, M. Callaghan, O. Edenhofer, et al. 2024. Systematic Review and Meta-Analysis of Ex-Post Evaluations on the Effectiveness of Carbon Pricing. *Nature Communications.* 16 May.

Edmonds. J. D. Forrister, L. Clarke, S. de Clara, and C. Munnings. 2019. *The Economic Potential of Article 6 of the Paris Agreement and Implementation Challenges.* International Emissions Trading Association and Carbon Pricing Leadership Coalition.

European Commission. 2024. EDGAR: Emissions Database for Global Atmospheric Research. GHG Emissions of all World Countries: 2023 Report.

Fang, Z. and Z. Jialu. 2023. State-Owned Enterprises' Responses to China's Carbon Neutrality Goals and Implications for Foreign Investors. *Georgetown Journal of International Affairs.* 15 February.

Fankhauser, S., C. Gennaioli, and M. Collins. 2015. The Political Economy of Passing Climate Change Legislation: Evidence from a Survey. *Global Environmental Change.* 35. pp. 52–61.

FAO (Food and Agriculture Organization of the United Nations), UNDP (United Nations Development Programme), and UNEP (United Nations Environment Programme). 2021. *A Multi-Billion-Dollar Opportunity: Repurposing Agricultural Support to Transform Food Systems.*

Flachsland, C., R. Marschinski, and O. Edenhofer. 2009. Global Trading versus Linking: Architectures for International Emissions Trading. *Energy Policy.* 37 (5). pp. 1637–1647.

G20. 2021. *G20 Rome Leaders' Declaration.*

G7. 2022. *G7 Leaders' Communiqué.*

Garg, V., B. Viswanathan, D. Narayanaswamy, C. Beaton, K., Ganesan, S. Sharma, and R. Bridle. 2020. *Mapping India's Energy Subsidies 2020: Fossil Fuels, Renewables, and Electric Vehicles*. International Institute for Sustainable Development.

Giardino, A., W. G. Wilson, and C. Hart. 2023. Decarbonizing the Water Sector. Background paper for the *Asia in the Global Transition to Net Zero: Asian Development Outlook Thematic Report 2023*. ADB.

Goulder, L. and M. Hafstead. 2013. Tax Reform and Environmental Policy: Options for Recycling Revenue from a Tax on Carbon Dioxide. *Resources for the Future Working Paper Series*. pp. 13–31.

Goulder, L. H., R. D. Morgenstern, C. Munnings, and J. Schreifels. 2017. China's National Carbon Dioxide Emission Trading System: An introduction. *Economics of Energy & Environmental Policy*. 6 (2). pp. 1–18.

Goulder, L. H., X. Long, J. Lu, and R. D. Morgenstern. 2021. China's Unconventional Nationwide CO_2 Emissions Trading System: Cost-Effectiveness and Distributional Impacts. *Journal of Environmental Economics and Management*. 111. 102561.

Grant, N., T. Aboumahboub, L. Welder, and C. Fyson. 2023. *Tripling Renewables by 2030: Interpreting the Global Goal at the Regional Level.* Climate Analytics.

Griscom, B., J. Adams, P. W. Ellis, and J. Fargione. 2017. Natural Climate Solutions. *Proceedings of the National Academy of Sciences*. 114. 44.

Gupta, S., S. Basu-Das, and S. Gupta. 2024. Carbon Pricing Through Emissions Trading In Asia-Pacific: A Comparative Analysis and the Way Forward. Background paper for the *Asia–Pacific Climate Report 2024: Catalyzing Finance and Policy Solutions*. ADB.

Haites, E. 2004. Tradeable Permits in the Policy Mix and Harmonisation of Emissions Trading Schemes. In OECD, ed. *Greenhouse Gas Emissions Trading and Project-Based Mechanisms*. Organisation for Economic Co-operation and Development. pp. 103–18.

Haites, E. 2015. Experience with Linking Greenhouse Gas Emissions Trading Systems. *WIREs Energy and Environment*. 5 (3). Wiley Interdisciplinary Reviews. pp. 246–60.

Han, M., F. Xie, and J. Zhang. 2023. Cross-Border Market Linkages in Asia and the Pacific: Theoretical and Empirical Underpinnings. Background paper for the *Asia–Pacific Climate Report 2024: Catalyzing Finance and Policy Solutions*. ADB.

ICAP (International Carbon Action Partnership). 2022. *New Zealand Emissions Trading Scheme*.

————. 2024. EU Emissions Trading System (accessed 10 July 2024).

ICAP (International Carbon Action Partnership) and World Bank. 2021. *Emissions Trading in Practice: A Handbook on Design and Implementation (2nd Edition).*

IEA (International Energy Agency). 2020. World Energy Investment 2020: Energy Financing and Funding.

————. 2023a. Government Energy Spending Tracker (July 2023 Update).

————. 2023b. World Energy Outlook 2023.

IGES (Institute for Global Environmental Strategies). 2024. Joint Crediting Mechanism (JCM) Database (accessed 4 August 2024).

IISD (International Institute for Sustainable Development) and OECD (Organisation for Economic Co-operation and Development). 2024. Fossil Fuel Subsidy Tracker (accessed 29 May 2024).

IPCC (Intergovernmental Panel on Climate Change). 2022. IPCC Sixth Assessment Report, Working Group III: Mitigation of Climate Change: Chapter 2: Emissions Trends and Drivers.

Jaffe, J. and R. Stavins. 2008. Linkage of Tradable Permit Systems in International Climate Policy Architecture. Working Paper. *NBER Working Paper Series.* National Bureau of Economic Research.

Jaumotte, F., W. Liu, and W. J. McKibbin. 2021. Mitigating Climate Change: Growth-Friendly Policies to Achieve Net Zero Emissions by 2050. *International Monetary Fund Working Paper.* No. 2021/195.

Konradt, M. and B. Weder di Mauro. 2021. Carbon Taxation and Greenflation: Evidence from Europe and Canada. *Journal of the European Economic Association.* 21 (6). pp. 2518–2546.

Köppl, A. and A. Schratzenstaller. 2023. Carbon Taxation: A Review of the Empirical Literature. *Journal of Economic Surveys.* 37. pp. 1353–1388.

Laan, T., A. Geddes, N. Do, L. Cameron, S. Goel, and N. Jones. 2023. Burning Billions: Record Public Money for Fossil Fuels Impeding Climate Action. *Energy Policy Tracker.*

Laan, T., A. Geddes, O. Bois von Kursk, N. Jones, K. Kuehne, L. Gerbase, C. O'Manique, D. Sharma, and L. Stockman. 2023. Fanning the Flames: G20 Provides Record Financial Support for Fossil Fuels. *Energy Policy Tracker.*

Landholm, D., F. Bravo, I. Palmegiani, S. Minoli, C. Streck, and S. Mikolajczyk. 2022. *Unlocking Nature Based Solutions Through Carbon Markets: Global Analysis of Available Supply Potential.* Climate Focus.

Li, M. and M. Duan. 2021. Exploring Linkage Opportunities for China's Emissions Trading System Under the Paris Targets – EU-China and Japan-Korea-China Cases. *Energy Economics.* 102. pp. 105–528.

Lu, C., Q. Tong, and X. Liu. 2010. The Impacts of Carbon Tax and Complementary Policies on Chinese Economy. *Energy Policy.* 38 (11). pp. 7278–7285.

Mansur, E. T. 2012. Upstream Versus Downstream Implementation of Climate Policy. In D. Fullerton and C. Wolfram, eds. *The Design and Implementation of US Climate Policy.* National Bureau of Economic Research.

McCulloch, N. 2023. *Ending Fossil Fuel Subsidies: The Politics of Saving the Planet.* Practical Action Publishing.

McKibbin, W., A. Morris, P. Wilcoxen, and Y. Cai. 2015. Carbon Taxes and U.S. Fiscal Reform. *National Tax Journal.* March. 68 (1). pp. 139–156.

McKibbin, W., R. Shackleton, and P. Wilcoxen. 1999. What to Expect from an International System of Tradable Permits for Carbon Emissions. *Resource and Energy Economics.* 21 (3). pp. 319–46.

Metcalf, G. and J. Stock. 2020. Measuring the Macroeconomic Impact of Carbon Taxes. *AEA Papers and Proceedings.* 110. pp. 101–06.

OECD (Organisation for Economic Co-operation and Development). 2023. Agricultural Policy Monitoring and Evaluation: Adapting Agriculture to Climate Change. 30 October.

PLN (Perusahaan Listrik Negara). 2021. Rencana Usaha Penyediaan Tenaga Listrik (RUPTL) PT PLN (Persero) 2021–2030.

Procton, A. 2024. *State of the Voluntary Carbon Market 2024: On the Pathway to Maturity*. Ecosystem Marketplace.

Qi, T. and Y. Weng. 2016. Economic Impacts of an International Carbon Market in Achieving the INDC Targets. *Energy*. 109. pp. 886–93.

Rahut, D., D. Narmandakh, X. He, Y. Yao, and S. Tian. 2023. Opportunities and Challenges During Low Carbon Transition: A Perspective from Asia's Agriculture Sector. Background paper for the *Asia in the Global Transition to Net Zero: Asian Development Outlook Thematic Report 2023*. ADB.

Ranson, M. and R. Stavins. 2016. Linkage of Greenhouse Gas Emissions Trading Systems: Learning from Experience. *Climate Policy*. 16 (3). pp. 284–300.

Raymond, L. 2020. Carbon Pricing and Economic Populism: The Case of Ontario. *Climate Policy*. 20 (9). pp. 1127–40.

Rennert, K., F. Errickson, B. C. Prest, L. Rennels, R. G. Newell, W. Pizer, C. Kingdon, J. Wingenroth, R. Cooke, B. Parthum, and D. Smith. 2022. Comprehensive Evidence Implies a Higher Social Cost of CO_2. *Nature*. 610 (7933). pp. 687–692.

Ricker-Gilbert, J. and T.S. Jayne. 2016. Estimating the Enduring Effects of Fertiliser Subsidies on Commercial Fertiliser Demand and Maize Production: Panel Data Evidence from Malawi. *Journal of Agricultural Economics*. 68.

Santikarn, M., L. Li, S. La Hoz Theuer, and C. Haug. 2018. A Guide to Linking Emissions Trading Systems. International Carbon Action Partnership.

Sgaravatti, G., S. Tagliapietra, C. Trasi, and G. Zachmann. 2023. National Fiscal Policy Responses to the Energy Crisis. Bruegel Datasets. 26 June.

Shapiro, A. F. and G. E. Metcalf. 2023. The Macroeconomic Effects of a Carbon Tax to Meet the Paris Agreement Target: The Role of Firm Creation and Technology Adoption. *Journal of Public Economics*. 218.

Tobing, L. 2022. How Indonesia's Pertamina Is Tackling the Energy Transition. Boston Consulting Group (BCG) Global. 27 May.

UNDESA (United Nations Department of Economic and Social Affairs). n.d. Goal 7: Ensure Access to Affordable, Reliable, Sustainable and Modern Energy for All.

UNEP (United Nations Environment Programme). 2023. *State of Finance for Nature: The Big Nature Turnaround, Repurposing $7 Trillion to Combat Nature Loss*.

_____. 2024. Copenhagen Climate Centre. Article 6 Pipeline (accessed 6 July 2024).

van Asselt, H., J. Kuehl, N. Jones, and J. Skovgaard. 2023. How the UNFCCC Can Tackle Fossil Fuel Subsidies at COP 28 and Beyond. International Institute for Sustainable Development. 27 November.

Viswanathan, B., S. Raizada, A. M. Bassi, G. Pallaske, and C. Beaton. 2022. *India's State-Owned Energy Enterprises, 2020–2050: Identifying Evidence-Based Diversification Strategies*. International Institute for Sustainable Development.

World Bank. 2015. *Carbon Leakage: Theory, Evidence and Policy Design.*

_____. 2016. *State and Trends of Carbon Pricing 2016* (Chapter 4).

_____. 2023a. *Building Public Support for Energy Subsidy Reforms*. 28 November.

_____. 2023b. *Detox Development: Repurposing Environmentally Harmful Subsidies.*

_____. 2023c. *State and Trends of Carbon Pricing 2023* (Chapter 2).

Yamazaki, A. 2017. Jobs and Climate Policy: Evidence from British Columbia's Revenue-Neutral Carbon Tax. *Journal of Environmental Economics and Management*. 83. pp. 197–216.

www.ingramcontent.com/pod-product-compliance
Lightning Source LLC
LaVergne TN
LVHW071447180726
843512LV00018B/1314